Working Together Apart:

Collaboration over the Internet

Synthesis Lectures on Human-Centered Infomatics

Editor
John M. Carroll, *Penn State University*

Human-Centered Informatics (HCI) is the intersection of the cultural, the social, the cognitive, and the aesthetic with computing and information technology. It encompasses a huge range of issues, theories, technologies, designs, tools, environments and human experiences in knowledge work, recreation and leisure activity, teaching and learning, and the potpourri of everyday life. The series will publish state-of-the-art syntheses, case studies, and tutorials in key areas. It will share the focus of leading international conferences in HCI.

Working Together Apart: Collaboration over the Internet
Judith S. Olson and Gary M. Olson
2013

Surface Computing and Collaborative Analysis Work
Judith Brown, Jeff Wilson, Stevenson Gossage, Chris Hack, Robert Biddle
2013

How We Cope with Digital Technology
Phil Turner
2013

Translating Euclid: Designing a Human-Centered Mathematics
Gerry Stahl
2013

Adaptive Interaction: A Utility Maximization Approach to Understanding Human Interaction with Technology
Stephen J. Payne and Andrew Howes
2013

Making Claims: Knowledge Design, Capture, and Sharing in HCI
D. Scott McCrickard
2012

HCI Theory: Classical, Modern, and Contemporary
Yvonne Rogers
2012

Activity Theory in HCI: Fundamentals and Reflections
Victor Kaptelinin and Bonnie Nardi
2012

Conceptual Models: Core to Good Design
Jeff Johnson and Austin Henderson
2011

Geographical Design: Spatial Cognition and Geographical Information Science
Stephen C. Hirtle
2011

User-Centered Agile Methods
Hugh Beyer
2010

Experience-Centered Design: Designers, Users, and Communities in Dialogue
PeterWright and John McCarthy
2010

Experience Design: Technology for All the Right Reasons
Marc Hassenzahl
2010

Designing and Evaluating Usable Technology in Industrial Research: Three Case Studies
Clare-Marie Karat and John Karat
2010

Interacting with Information
Ann Blandford and Simon Attfield
2010

Designing for User Engagement: Aesthetic and Attractive User Interfaces
Alistair Sutcliffe
2009

Context-Aware Mobile Computing: Affordances of Space, Social Awareness, and Social

Influence
Geri Gay
2009

Studies of Work and the Workplace in HCI: Concepts and Techniques
Graham Button and Wes Sharrock
2009

Semiotic Engineering Methods for Scientific Research in HCI
Clarisse Sieckenius de Souza and Carla Faria Leitão
2009

Common Ground in Electronically Mediated Conversation
Andrew Monk
2008

Copyright © 2014 by Morgan & Claypool

All rights reserved. No part of this publication may be reproduced, stored in a retrieval system, or transmitted in any form or by any means—electronic, mechanical, photocopy, recording, or any other except for brief quotations in printed reviews, without the prior permission of the publisher.

Working Together Apart: Collaboration over the Internet
Judith S. Olson and Gary M. Olson

www.morganclaypool.com

ISBN: 9781608450503 print
ISBN: 9781608450510 ebook

DOI 0.2200/S00542ED1V01Y201310HCI020

A Publication in the Morgan & Claypool Publishers series
SYNTHESIS LECTURES ON HUMAN-CENTERED INFORMATICS #20
Series Editor: John M. Carroll, Penn State University

Series ISSN 1946-7680 Print 1946-7699 Electronic

Working Together Apart:
Collaboration over the Internet

Judith S. Olson and Gary M. Olson
University of California, Irvine

SYNTHESIS LECTURES ON HUMAN-CENTERED INFORMATICS #20

MORGAN & CLAYPOOL PUBLISHERS

ABSTRACT

Increasingly, teams are working together when they are not in the same location, even though there are many challenges to doing so successfully. Here we review the latest insights into these matters, guided by a framework that we have developed during two decades of research on this topic.

This framework organizes a series of factors that we have found to differentiate between successful and unsuccessful distributed collaborations. We then review the kinds of technology options that are available today, focusing more on types of technologies rather than specific instances. We describe a database of geographically distributed projects we have studied and introduce the Collaboration Success Wizard, an online tool for assessing past, present, or planned distributed collaborations. We close with a set of recommendations for individuals, managers, and those higher in the organizations who wish to support distance work.

KEYWORDS

distance work, virtual teams, teamwork, distributed teams, managing virtual teams, communication, coordination, technology support, infrastructure, cyberinfrastructure, team science, trust

Contents

Acknowledgments xiii

1   **The Changing Landscape** 1

2   **Types of Distributed Collaborations** 7
    2.1   Distributed Project or Enterprise 8
    2.2   Shared Instrument or Resource 9
    2.3   Community Data Bases 12
    2.4   Open Contribution System 14
    2.5   Virtual Community of Practice 16
    2.6   Virtual Learning Community 17
    2.7   Community Infrastructure Project 18
    2.8   Remote Expertise 19
    2.9   Evolution from One Type to Another 20
    2.10   Some Other Factors 21
    2.11   Summary 22
    2.12   Key Attributes 23
    2.13   Using the Typology 23

3   **What It Means To Be Successful** 25
    3.1   Success in Research: The Sciences and the Humanities 25
        3.1.1   Ideas 26
        3.1.2   Tools 27
        3.1.3   Training 27
        3.1.4   Outreach 28
        3.1.5   Impact 29
    3.2   Success in Corporations 30
    3.3   Success in Non-Profits 32
    3.4   Summary of Successes 32

4   **Overview of Factors that Lead to Success** 33

5   **The Nature of the Work** 35

6   **Common Ground** 39

7	**Collaboration Readiness**	43
	7.1 Characteristics of the Individual Team Members	43
	7.2 The Culture of Collaboration or Competition	44
	7.3 Examination of the Explicit Sources of Motivation	44
	7.4 Trust	46
	7.5 Group Self-Efficacy	47
8	**Organization and Management**	49
	8.1 The Project Organization	49
	8.2 The Project Manager	49
	8.3 What's Special about Managing Distributed Work?	50
	8.4 What Management Includes	51
	8.4.1 Plans	51
	8.4.2 Decision Making	52
	8.4.3 Managing Across Time Zones and Cultures	52
	8.4.4 Managing Legal Issues	53
	8.4.5 Managing Financial Issues	54
	8.4.6 Managing Knowledge	54
	8.4.7 Launching a Distributed Project	55
	8.5 Summary	55
9	**Collaboration Technologies and Their Use**	57
	9.1 Kinds of Collaboration Technologies	58
	9.1.1 Communication Tools	58
	9.1.2 Coordination Tools	66
	9.1.3 Information Repositories	70
	9.1.4 Computational Infrastructure	72
	9.2 Deciding What Constellation of Technologies a Particular Collaboration Needs	75
	9.2.1 Speed	76
	9.2.2 Size	77
	9.2.3 Security	77
	9.2.4 Privacy	78
	9.2.5 Accessibility	78
	9.2.6 Control	78
	9.2.7 Media Richness	80
	9.2.8 Ease of Use	80
	9.2.9 Context Information	82

		9.2.10	Cost	83
		9.2.11	Compatibility with Other Things Used	83
	9.3	Example Decisions about Technology Choices		83
	9.4	Conclusions		85
10	**The Science of Collaboratories Database**			**87**
	10.1	Information Collected		87
	10.2	Findings to Date		91
11	**The Collaboration Success Wizard**			**97**
	11.1	Details of the Wizard		97
	11.2	Details of the Reports		101
	11.3	Initial Experience with the Wizard		102
	11.4	The Wizard as Translational Research		103
	11.5	Contact Re the Wizard		104
12	**Summary and Recommendations**			**105**
	12.1	What about Distance Matters?		106
		12.1.1	Blind and Invisible	106
		12.1.2	Time Zone Differences	107
		12.1.3	Crossing Institutional or Cultural Boundaries	107
		12.1.4	Uneven Distribution and the Consequent Imbalance of Power or Status	107
	12.2	Recommendations		107
	12.3	Recommendations Concerning the Individuals Who are Members of a Distributed Team		108
		12.3.1	Collaboration Readiness	108
		12.3.2	Technical Readiness	108
	12.4	Recommendations for the Manager of a Distributed Team		108
		12.4.1	Selecting People for the Team	108
		12.4.2	Common Ground	109
		12.4.3	Collaboration Readiness	109
		12.4.4	The Nature of the Work	110
		12.4.5	Management	110
		12.4.6	Technology Readiness	111
	12.5	Recommendations for an Organization that Wishes to Support a Distributed Team		111
	12.6	In the Future, Will Distance Still Matter?		112

References 115

Author Biographies 137

Acknowledgments

We have been studying long-distance collaboration since the mid-1980s, when we got involved in the National Science Foundation's EXPRES Project (Olson and Atkins, 1990). The National Science Foundation has continued to be the primary source of funding for our work in this area, through a series of projects (ASC-8617699, IRI-8902930, IRI-9216848, IRI-9320543, ATM-9873025, IIS-9977923, IIS-0085951, CMS-0117853, IIS-0308009, OCI-1025769, ACI-1322304). However, we've also had support from the Army Research Institute (W74V8H-06-P-0518, W91WAW-07-C-0060), Google, U.S. West, Anderson Consulting, Apple, Intel, Ameritech, and the John D. Evans Foundation. While we have come at the issues of long-distance work primarily from the perspective of the field of Computer Supported Cooperative Work (CSCW), which we recently reviewed in (Hall et al. 2008), we have found that many other fields have looked into issues of such work. To help us with this we have had the valuable input from our multidisciplinary colleagues at the School of Information at the University of Michigan and the Department of Informatics at the University of California, Irvine. Those colleagues who have particularly influenced us include Mark Ackerman, Dan Atkins, Geof Bowker, Yan Chen, Bob Clauer, Derrick Cogburn, the late Michael Cohen, Paul Dourish, Tom Finholt, George Furnas, Joseph Hardin, Tim Killeen, Jeff Mackie-Mason, Gloria Mark, and Stephanie Teasley. One major influence on our thinking about these matters came from a challenge first presented to us by Suzi Iacono from the National Science Foundation, who asked at a conference in Vienna: Why do some long-distance collaborations work, and some not? We subsequently took up this challenge initially for scientific collaboration, in the Science of Collaboratories (SOC) Project, funded by the National Science Foundation. Later, we expanded this to include corporate settings with sponsorship from the Ford Motor Company. The early form of the SOC project was informed by an external advisory committee that included Mark Ellisman, Jim Herbsleb, Jim Myers, Diane Sonnenwald, and Nestor Zaluzec. With the advice of this group, we held a series of workshops, some in Ann Arbor and some in Washington, D.C. Summaries of these workshops, including all the people who participated in them, can be found at soc.ics.uci.edu/workshops/. Many people have worked with us over the years, many of who appear as coauthors on papers we have cited in this book. But several need to be mentioned explicitly for the major roles they had in shaping our thinking and guiding our work: Matthew Bietz, Nathan Bos, Dan Cooney, Erik Hofer, Airong Luo, Emily Navarro, Sue Schuon, Ann Verhey-Henke, Amy Voida, Jude Yew, Ann Zimmerman, and the late Steve Abrams. Several people have read parts or this entire book in draft form, including the late Steve Abrams, Matt Bietz, Jack Carroll, and Jonathan Cummings.

CHAPTER 1

The Changing Landscape

The rise of a widespread, reliable, and high-speed Internet has enabled groups to work together successfully when they are not in the same geographic location. Allen (1977) reported that the likelihood of interacting with another person falls off rapidly with distance and essentially asymptotes at 30 m. This means that people in the same building but in different wings or different floors or in different buildings on the same campus need to rely on these emerging distance technologies just as much as those in different cities, states, or countries. And the increase in this kind of work is incredible, no matter which domain of activity we look at.

The figures on the extent of virtual teams in contemporary corporations are staggering. The Institute for Corporate Productivity[1] stated in 2008 that 67% of the companies that were surveyed felt that "their reliance on virtual teams" would grow in the next few years.[2] For companies that had more than 10,000 employees, this figure was more than 80%. Companies are distributed in order to reach new markets, gain access to specialized resources and expertise, and/or change the costs of doing their work. As Thomas Malone (2004) notes, these new technologies not only enable distributed work but they change the very nature of how enterprises are organized and carry out their mission.

The worlds of research and scholarship have changed equally dramatically. In the 1990s in the U.S., within the National Science Foundation, a new form of collaboration in science received great attention, dubbed the Collaboratory (Wulf, 1993), a laboratory without walls. A parallel form in the United Kingdom is called variously eScience or eResearch (Jankowski, 2009). Collaboratories and eScience arose because many problems in science and engineering are large and complex. No one university houses sufficient numbers of experts in a field, requiring collaboration across distance. For instance, in the physical sciences, instrumentation is increasingly expensive and therefore must be shared. The Large Hadron Collider in high energy physics is a contemporary and classic example (Hofer et al., 2008), and follows a pattern in this field since the Manhattan Project during the 1940s. Fields as diverse as upper atmospheric physics (Olson et al., 2008), earthquake engineering (Spencer et al., 2008), and environmental molecular science (Myers, 2008), all require sharing of highly specialized, expensive equipment. In many sciences the creation of large databases is a key next step in advancing. Examples include GenBank (Pevsner, 2009), the Protein Data Bank (Bernstein et al., 1977; Berman et al., 2003), National Virtual Observatory (NVO) (Ackerman et

---

[1] http://www.pr.com/press-release/103409.
[2] Throughout this book we are agnostic about the actual physical location of the distributed participants. The growing trend to work from home is one example of how participants might be distributed. While there may be special issues for this particular location, we will not focus on that here.

al., 2008), the Biomedical Informatics Research Network (BIRN) (Olson et al., 2008), and the Long-Term Ecological Research (LTER) program (Michener and Waide, 2008). By coordinating their work through distance technologies or working with remote instruments, aggregated data or computing resources, academics can attack bigger questions with the promise of breakthroughs in our understanding and in solving problems. The extent of this revolution is documented in Chapter 10, where we note that our Science of Collaboratories Database has grown to more than 700 instances of such projects, even though this used a less-than-efficient opportunistic sampling strategy. Science and engineering are forever changed.

This trend to large-scale distance work is extending to domains in the social sciences and humanities, in fields that have been historically slower to pick up the strategies of collaborating on research and publishing multi-authored papers and books. People collaborate across universities to visualize social processes, do computer modeling, simulate social and political networking, and examine consequences of natural events, such as how rising water from global warming will affect low-lying London (Borning et al., 2005). In the humanities there are virtual haptic museums, and vast collections of material such as all the written material in Greek from ancient times to the fall of Constantinople, allowing researchers to find, sort, and comment on findings electronically (Inman et al., 2004) .

Many non-profits, like the Red Cross or Red Crescent, or the Girl or Boy Scouts, have been distributed for a long time, but now that there are better communication and coordination tools, they can be more effective. Distance communication has helped with the sharing of best practices, has facilitated fund raising, and, more recently, and has led to more rapid responses to disasters. For example, researchers have found that people *in* a natural disaster can help each other and their rescuers through reporting what is happening in their location through microblogging, such as Twitter (Palen and Liu, 2007; Palen et al., 2011). However, there are similar challenges in distributed non-profits as there are in academia and corporations. There are issues of focus, culture, buy-in, and trust (Lewis et al., 2010).

The evidence of increased collaboration is striking. For instance, Wuchty, et al. (2007) looked at publications and patents, and found in a broad range of areas larger teams were either publishing papers or acquiring patents. Such indices are found for publishing data in many fields (Porter and Rafols, 2009). Page (2007) points out similar pattern even for Nobel Prizes. The number of awardees per Nobel Prize has risen from 1.2 per prize in 1901 to 2.75 in 2004 in both Chemistry and Physics.

In addition to the challenges of being distributed, more and more teams today are multi-disciplinary. Multi-disciplinary teams are deemed necessary in order to attack complex problems. Chamberlin (1890) long ago argued for bringing multiple perspectives to bear on scientific problems. More recently, Page (2007) described, in great detail, the potential advantages of having diverse conceptual perspectives and problem-solving strategies in coming up with insights and solutions to difficult problems. Stokols et al. (2010) describe the many incentives for carrying out research in the biomedical sciences using teams that span disciplines. Research projects that focus on a real world problem, like disease, energy, the economy or the environment, tend to bring together individuals from many disciplines. For example, in studying the structure of the brain (a mouse brain, a good model of the human brain for some purposes), the Morphometry Biomedical Informatics Research Network (mBIRN) brought together researchers spanning from the molecular structure in the brain all the way up to its morphometry. Furthermore, these individuals are almost never collocated. The Atlas Project of the Large Hadron Collider at CERN in Geneva, Switzerland, involves 3000 physicists from 38 different countries, 174 universities and laboratories, and 1000 students (see Figure 1.1). A project of this scale would not be possible without contemporary computing and communication technologies.

Figure 1.1. A meeting of the ATLAS collaboration.

While the growth of such distributed projects testifies to the success that has characterized many projects, we know that both distance and the fact that the teams are multi-disciplinary create formidable challenges. Despite the availability of increasingly sophisticated cyberinfrastructure to support all of this activity, working at a distance is still difficult. Individuals from different disciplines working together suffer from lack of common vocabulary and working styles. Olson and Olson (2000) reviewed data from studies of both scientific projects and corporate teams, and documented many of the problems. The Science of Collaboratories (SOC) Project, on which much of this volume is based, began formally just as that article was published. A lot has happened in the past decade and a half. This book is our compilation of what is known today about the challenges of working remotely and with people from multiple disciplines. It additionally offers actions that can be taken to mitigate those challenges. We recently published a book on the SOC Project, *Scientific Collaboration on the Internet*, that extended our earlier work and included illustrations with many case studies (Olson et al., 2008). Cummings and Kiesler (2005; 2007; 2008) carried out a series of quantitative studies of large collaborative projects at NSF and reported many insights into the key issues involved. A particularly fruitful line of work has developed under the heading of the Science of Team Science (SciTS). An annual conference series was launched in 2010 (a brief report appeared in Falk-Krzesinski et al., 2011). A special issue of the *American Journal of Preventative Medicine* on SciTS appeared in 2008 (Stokols et al., 2008, introduced the special issue). A group at the National Institutes of Health (NIH) has produced a report entitled "Collaboration and Team Science: A Field Guide" (Bennett et al., 2010), and a website rich with resources is at teamscience.nih.gov. Falk-Krzesinski, Borner et al. (2011) conducted a concept-mapping study to attempt to give some structure to this emerging field.

Similarly, one indicator of the concern for managing virtual and multi-disciplinary teams in the corporate world is the huge number of books published on how to create and manage effective teams (some recent examples are: Hackman, 2002, 2011; Sawyer, 2007; Hansen, 2009; Rosen, 2009). Business schools, engineering colleges, and information schools have added a variety of team projects to their curricula to prepare students for a world in which teamwork is common, valued, and challenging.

Our goal in this book is to bring together many of these threads, building on our initial work in the SOC project but augmenting it with insights emerging from this wider range of investigations. In the Chapter 2 we will first explore the different types of distance collaborations, each form of which encounters some special challenges. We present examples of each type in the for-profit and non-profit worlds as well as in the science and the humanities. Next, in Chapter 3 we cover what it means to be successful in collaborations, for there can be success in different facets as well as at different levels. We follow in Chapter 4 with a brief overview of the factors that we have found to lead to success, and then, in Chapter 5–9, a chapter for each major category with details and examples of each. We close with two chapters of practical import. Chapter 10 describes an on-line resource for researchers in this area, the Science of Collaboratories database; Chapter 11 introduces an online assessment tool, the Collaboration Success Wizard, that embodies the theory presented in this book. We end in Chapter 12 with a summary of the findings and a list of recommendations for individuals, managers, and organizations involved in distance work

> **Note:** Although the challenges and remedies reported in this book apply to the corporate world and non-profits as well as academic collaborations, most of our experience has been with scientific collaborations, the "Collaboratories" described above, and thus a majority of our examples are from them.

CHAPTER 2
Types of Distributed Collaborations

Not all distance collaborations are the same. Those that are similar might face similar issues, and by classifying them we might be better able to find best practices of the similar issues they face.

Beginning in 2001, we collected examples of science collaboratories using a variety of leads to find examples of such collaboratories, called a purposeful sample (Patten, 1990). This included notes from talks, mentions in news reports, links in various websites, personal contacts, and the technical literature in numerous fields. We collected basic facts about each and entered them into a standard form in a database, which we will describe in more detail in Chapter 10. The taxonomy first described in Bos et al. (2008) was developed by a process of finding similarities and differences in the projects' structure and purpose. For many, we noted a primary type and a secondary or even tertiary type. Many collaborations change character over time, a point we'll return to at the end of this chapter. Over the last ten years, we continued to find new (to us) collaborations, first in other academic areas, primarily humanities and the social sciences, and then in the corporate and non-profit worlds. In this chapter we present an updated taxonomy that reflects these new findings.

In what follows, for each type, we give a definition, some examples from a variety of settings, the key kinds of technology and social issues in this type, and key organizational issues.

Table 2.1: Eight types of distributed collaboration

Name	Definition
Distributed Project or Enterprise	Aggregated talent, effort, and resources with a common purpose
Shared Instrument or Resource	Remote access to an expensive or rare instrument, or a resource such as high-end computation
Community Data Bases	A database that is created, maintained, or improved by a geographically distributed community
Open Community Contribution System	An open project that aggregates the efforts of many geographically separate individuals through microcontributions
Virtual Community of Practice	A network of individuals who share an area of interest and communicate about it online

Virtual Learning Community	A network of individuals who have banded together to jointly learn a rare skill or topic
Community Infrastructure Project	A distributed community that builds the infrastructure and tools to collaborate
Remote Expertise	Access to problem solving from a remote person

2.1 DISTRIBUTED PROJECT OR ENTERPRISE

This is the quintessential distributed work, where people in different locations share a common goal and aggregate their talent, effort and/or resources. This form of working is becoming more and more common across all domains. In our Science of Collaboratories database (see Chapter 10), we have now identified over 700 such projects. In *The Future of Work* (Malone, 2004) and many other recent writings geared toward the world of management, such virtual enterprises are described as the direction the world is going (e.g., Thompson, 2008). Communication is at the heart of effective organizations, and as the cost of communication declines through modern information technology, the possibilities for successful distributed organizations increase. Voida et al. (2011) found similar trends in the world of non-profits.

In the world of science, the *Alliance for Cellular Signaling* (www.afcs.org) exemplifies this kind of distributed work. Scientists from eight U.S. universities worked to create a complete model cell, trying to account for the cell signaling behavior (Abbott, 2002). Results of this and related projects can been found at The *Signaling Gateway* at UCSD, where the key collaborative artifacts, namely Molecule Pages, are maintained in a large data base (www.signaling-gateway.org). We will return to aspects of this database later in this chapter. In the humanities, de la Flor et al. (2010) describe a virtual research environment (VRE) that supported the geographically dispersed collaborative analysis of ancient texts. In social science, for example, MoSeS (www.ncess.ac.uk/research/geographic/moses/) applied modeling and simulation to policy-making in reference to health, business and transport. Dutton and Meyer (2010) review a number of other similar collaborations in the social sciences. In the non-profit world, this type is exemplified by Red Cross disaster relief and Susan G. Komen Foundation's coordination of various fundraising efforts.

Since this type of collaboration is attempting to mimic face-to-face collaboration, technologies that allow awareness of other participants and easy, expressive communication are central. An early humanities collaboration, called *Imaging Florida*, used Netmeeting's video, instant messaging and email in rich combination (Freeman, 2004). If, in addition, there are data being collected and shared, or an instrument is being shared, the kinds of technical issues in Sections 2.2 and 2.3 are relevant.

Similarly, organizational issues abound in this type of collaboration. Participants need to stay motivated, which is difficult when you do not bump into your colleagues or note their working. Collaborators must work to standardize their contributions; they need to create procedures for decision making and administrative support. In academics and the corporate world, there are issues of intellectual property that get messy when you cross institutional lines. And there are career development issues when one's evaluator/manager is not collocated with the person and is "out of sight, out of mind" (Hemphill and Begel, 2008; Koehne et al., 2012).

2.2 SHARED INSTRUMENT OR RESOURCE

In science, shared instrument collaborations involve unusually expensive instruments that are beyond the reach of any individual project. These are then set up to allow many researchers to access them remotely. For example, the Environmental Molecular Sciences Laboratory (EMSL) makes its nuclear magnetic resonance (NMR) instruments available to scientists and engineers who publish their research openly (Myers 2008). Special terms and conditions apply if the research is proprietary. The Upper Atmospheric Research Collaboratory (UARC) began as a resource for upper atmospheric physicists to access remote instruments on the west coast of Greenland (Olson et al., 2008) but grew to share instruments around the world. The NMR at the Beckman Institute at the University of Illinois could be accessed from classrooms to examine chicken embryo, in a project called *Chickscope* (chickscope.itg.uiuc.edu). In the humanities, academics can access a virtual archeology environment through *Ceren*, the name of a village in El Salvador that, like Pompeii, was buried in ashes in the 7th century (ceren.colorado.edu). In the corporate world, Google and Nintendo give access to developers to a testing portal to run tests on their Web applications or Nintendo games.

Another kind of shared resource is access to high end computational resources, such as supercomputers or massive clusters. Such facilities are needed when huge amounts of data need to be analyzed, such as those that arise in fields like high energy physics, fluid dynamics, weather and climate modeling, astrophysics, molecular biology, and many other fields. One particular subset of these needs is the creation of scientific visualizations. Large, high-definition displays make it possible to create visualizations of enormous complexity and detail, including a variety of three-dimensional displays (See Figure 2.1).

One current example of the need to process huge amounts of data comes from the Large Hadron Collider (LHC) at the European Center for Nuclear Research (CERN) in Geneva, Switzerland. When in operation the LHC produces about 15 petabytes of data per year. As a result, a large networking and computing infrastructure was created to allow physicists around the world to analyze data from the LHC experiments. A description of the LHC and how such infrastructure was created is in Hofer et al. (2008).

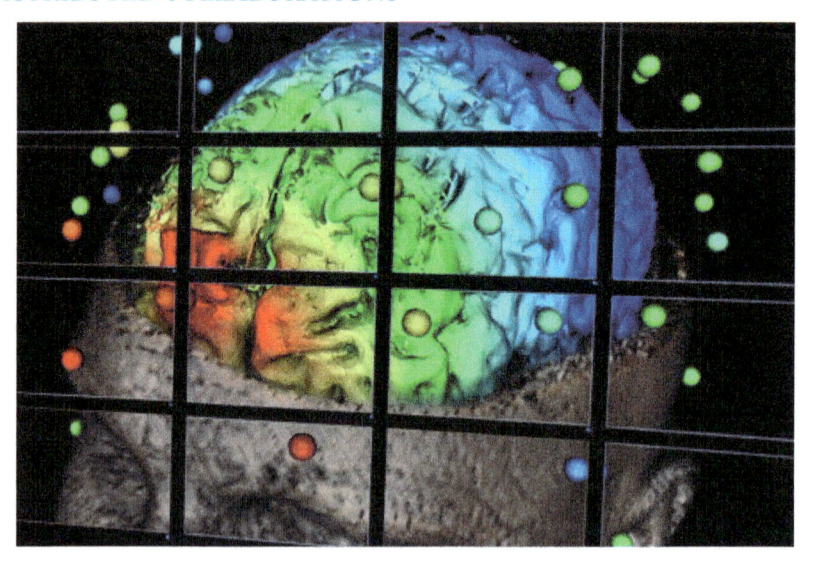

Figure 2.1: High-resolution, wall-sized visualization of the human brain.

One particularly interesting example of the use of high end computing was carried out in the early phase of the George E. Brown, Jr., Network for Earthquake Engineering Simulation (NEES), a large-scale, long-term project funded by NSF. In July 2003 the NEES project carried out a Multi-Site Online Simulation Test (MOST). This test consisted of mixing together physical and computational simulation in the same experiment. The engineering scenario, illustrated in Figure 2.2, was to have a simple physical structure that would have a force projected on one end (the Moment in Figure 2.2) that would be propagated through a steel structure and the results measured at other points (the Pinections in Figure 2.2). The interesting wrinkle was that the two ends of this structure were physically implemented at the University of Illinois and the University of Colorado, while the middle portion was computationally implemented at the National Center for Supercomputing Applications in Champaign, Illinois. Figure 2.2 shows the arrangement. From Hofer, et al.: High-energy physics: The large hadron collider collaborations. Scientific Collaboration on the Internet, pages 143-151. G. M. Olson, A. Zimmerman, and N. Bos. Cambridge, MA: MIT Press. Copyright © 2008 Massachusetts Institute of Technology. Used with permission.

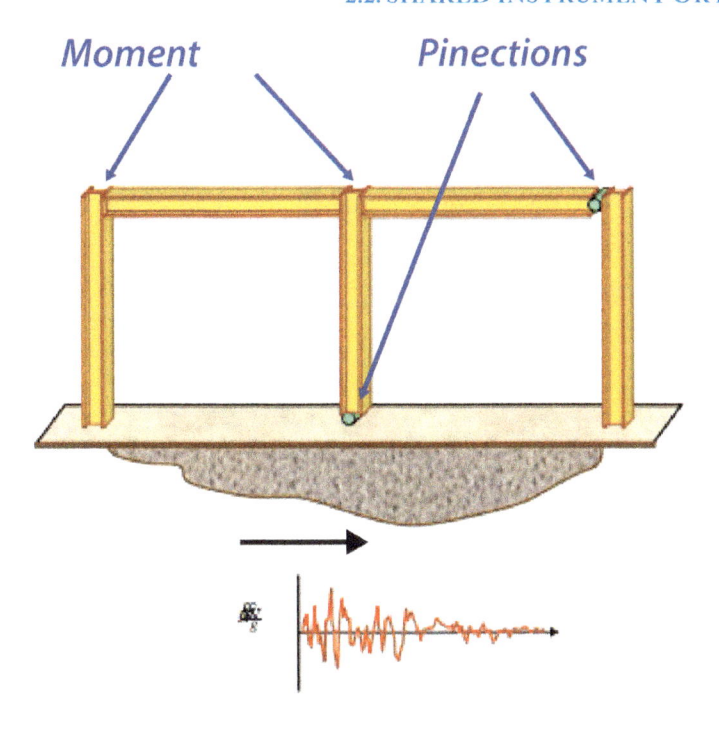

Figure 2.2: The steel frame structure that was modeled in the NEES MOST experiment.

The MOST experiment was successful, and demonstrated for the earthquake engineering community that a useful hybrid strategy could be employed in carrying out tests of the stresses on physical structures that result from motions that simulate potential earthquake motions. Technical details of the MOST experiment are available in a technical report at www.neesgrid.org/most/. The most recent information about the NEES project, now known as NEEShub, is available at nees.org. Of particular interest is information about the current use computational models at nees.org/simulation.

12   2. TYPES OF DISTRIBUTED COLLABORATIONS

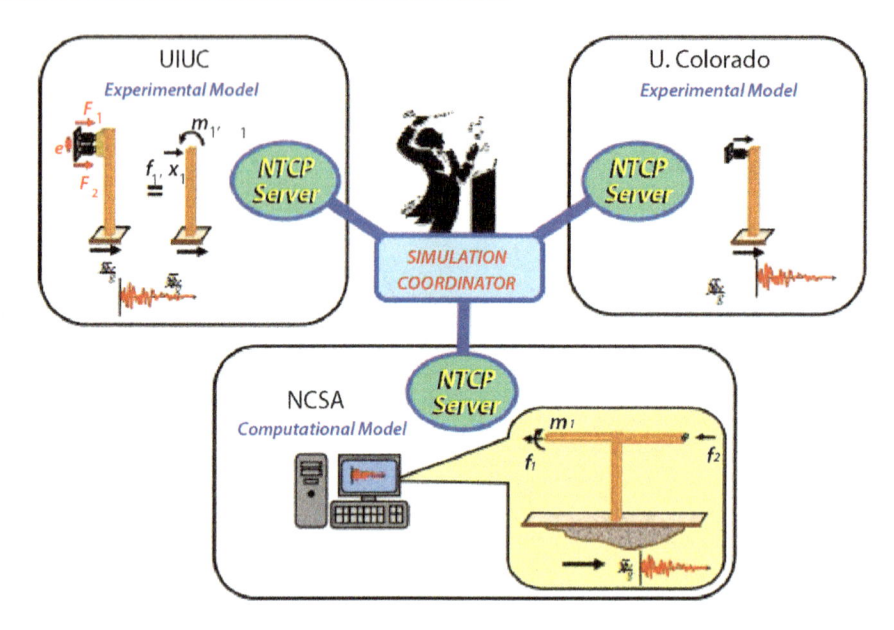

Figure 2.3: The three components of the NEES MOST experiment. Adapted from http://nees.org/.

Technical issues abound in shared instruments and resources. Bandwidth needs to be sufficient for the data transfer and delays are unacceptable when controlling an instrument in real time. There are also safety issues when remote instruments are being controlled by someone at a distant site. As mentioned, some data flows are monumental, and in some cases even in the first stage the data must be sampled before it is shared over a network. The data transfer and storage requirements are huge, though thankfully, there has been rapid development of advanced networking and storage capabilities (e.g., Tanenbaum and Wetherall, 2011). Access issues and data security are important issues as well.

Organizational issues surround procedures for allowing access, which involves difficult choices when many people want to use them. Many organizations develop committees to oversee this process and to ensure fair distribution of time. There are also issues of certification of users, often highlighting different customs in different cultures. For example, in Japan, the education needed for running a large collider is only at the bachelors level, whereas in the U.S., it requires a Ph.D. Discussions of the organizational issues that arise in large-scale projects are found in Spencer et al. (2008) for NEES and Hofer et al. (2008) for LHC.

2.3   COMMUNITY DATA BASES

Data and/or information are at the heart of contemporary collaborations, whether in science, engineering, or business. The 2003 National Science Foundation report on cyberinfrastructure (Atkins et al., 2003) reviewed in considerable detail the many issues surrounding data in scientific research, including matters like standardization, curation, access control, confederation, metadata, etc. Also, working with data also entails tools for processing and visualizing, and training people to use these tools. These matters get especially complex when members of different disciplines collaboration on cutting edge problems. Zimmerman (2008) reports examples from ecological research that data collected for different purposes or by researchers with different skills can affect their reusability.

In the business world, the enormous interest in knowledge management (e.g., Hislop, 2013) is an indication of how valuable information is in the contemporary enterprise. The many complexities surrounding the introduction of electronic medical records (e.g., Park et al., 2012) reveal how important data issues are in the medical realm. Voida and her colleagues (Voida et al., 2011) underscored the importance of data management in the non-profit world. Controversies about information and data security abound as a lot of infrastructure migrates to the "cloud" (Marshall and Tang, 2012; Voida et al., 2013) further underscoring the sensitivity and importance of data management issues.

Not surprisingly, then, many of the geographically distributed projects that we studied have had the development and maintenance of data repositories as their central organizing goal. For example, a number of recent projects in science and engineering collaborate to standardize their procedures and share their data. One of the projects in the Biomedical Informatics Research Network (BIRN) involves collecting standardized data on functional images of the brains of schizophrenics (Olson et al., 2008). By standardizing the equipment and the protocol and sharing their data in a shared data repository, science can advance much more quickly because of the larger sample size. In the humanities, the *Thesaurus Linguae Graecae* project has digitized most of the literary texts written in Greek, starting with Homer and up to the fall of Constantinople in 1453 (www.tlg.uci.edu). This resource is a collection of digital documents and their translations, today containing 10,000 works.

In many corporations, knowledge bases are contributed to by people over time. Dow Chemical, for example, has a repository of all the laboratory tests it has ever conducted, an impressive archive that prevents them from having to redo expensive tests. In the non-profit world, an example is the national list of political donors (www.fec.gov/finance/disclosure/norindsea.shtml) and the Red Cross database that local chapters contribute to noting their local volunteers and what skills they have been certified in (www.redcross.org/find-your-local-chapter).

Community Data Bases often lead standardization efforts. Sometimes it takes agreement from many disparate parties to form standards, but these standards are essential in allowing people to aggregate their data and store it in a searchable format. In many science communities, in particular, there are a number of technical tools developed to model and visualize aspects of the data. An example is the d3 environment, which is a JavaScript library for developing visualizations (d3js.org). Visualizations are a powerful method for understanding data (Card, 2012; Ware, 2013).

An important organizational issue is how to motivate individuals to contribute (Bos, 2008). Some science collaboratories have developed policies that make it mandatory to contribute data if a person wishes to use the repository. Some journals that publish studies based on new gene sequences require that the sequence data be submitted to *GenBank* before you can publish your work. In order to protect the author(s); rights to their findings, GenBank will hold off releasing the sequence data until a declared publication date.[3] In the corporate world, the dilemma in competitive environments like consulting companies is how to motivate individuals to contribute their best ideas and practices because it is by their ideas that they are judged for promotion or partnership (Orlikowski, 1992).

A recent phenomenon with huge implications for Community Data Systems is the emergence of what is being called "Big Data." While large data sets have been the norm in many areas of science for a long time, the emergence of sensory technology, logging of software activity, and in general the large scale of data capture of human activity has led to the creation of gigantic data sets that create all manner of commercial and scientific opportunity. Sites such as Google or Facebook make extensive use of such data to tailor aspects of the user experience. Amazon and others use large amounts of data to track user behavior and make recommendations. And social scientists are intrigued with the opportunities for studying behavior in a richness of detail over time that opens the door to new forms of inquiry. Of course, the phenomenon of Big Data has led to all manner of concerns about accuracy, privacy, data access and ownership, and opportunities for mischief. The recent revelations of the extent to which the U.S. government has been mining data in an alleged desire to forestall terrorists is only the most dramatic example of these concerns.

2.4 OPEN CONTRIBUTION SYSTEM

One of the most exciting recent developments is the facilitation via digitization of an activity that goes by a number of different names: citizen science (Bonney et al., 2009; Hand, 2010), the wisdom of crowds (Surowiecki, 2005; Shirky, 2008), collective intelligence (Malone et al., 2010), smart mobs (Rheingold, 2002), and crowdsourcing (Howe, 2008; Brabham, 2013). The idea that something very intelligent and useful can arise through a series of microcontributions (Sproull and Kiesler, 2005) is actually quite old, and indeed is a key element of the underlying mechanics of market economies. Supposedly, Francis Galton was surprised that the average guess of the weight of an ox calculated across the individual estimates of the members of a crowd was very close to accurate, whereas the individual estimates were mostly far off. This is close to the standard demonstration often done in introductory economics classes where members of the class are asked to estimate the number of pennies in a jar. Once again, the average of the class's estimates is usually very close to the actual number of pennies, even though the individual estimates show a wide dispersion.

---

[3] See details of this policy at www.ncbi.nlm.nig.gov/genbank/.

2.4. OPEN CONTRIBUTION SYSTEM 15

Surowiecki (2005) listed the criteria that need to be in place for the wisdom of crowds to emerge: diversity of opinion, independence of opinions, decentralization, and a mechanism for aggregating the individual private judgments into a collective decision. By a similar analysis, conditions that lead to a failure of collective wisdom can be listed: homogeneity of opinions, centralization of decision making, division or isolation of opinions, imitation, and emotionality (Surowiecki, 2005). Another key element is that each individual's judgment is modest in size, a microcontribution as noted above.

The Internet and subsequent digitization of tasks has yielded a large number of successful projects along these lines. For example, the *Clickworkers* project had public volunteers look for craters on images of the surface of Mars from the Viking Orbiter. The statistically aggregated inputs of such volunteers was indistinguishable from the judgments of experts (Szpir, 2002). *Galaxy Zoo* has citizens differentiate two kinds of galaxies from images gathered by the Sloan Digital Sky Survey (Lintott et al., 2008). Interestingly, a teacher in the Netherlands, Hanny van Arkel (Figure 2.4), discovered a third kind of galaxy while participating in the Galaxy Zoo project, dubbed Hanny's Voorwerp (Dutch for "object"; Wiggins and Crowston, 2010). *FoldIt* has citizens using their remarkable perceptual skills to figure out how to fold proteins (Cooper et al., 2010); in a recent breakthrough, citizens working in FoldIt were able to determine the structure of a key protein associated with the HIV virus (Khatib et al., 2011). Hand (2010) reviews a number of other such projects.

Figure 2.4: Hanny van Arkel, discoverer of Hanny's Voorwerp. Copyright © Adrie Mouthaan. Used with permission.

Similar projects occur outside of the science domain, and provide some useful insights into the mechanisms for such projects. Luis von Ahn invented a guessing game called the ESP game in which individuals, paired with another randomly chosen person on the Internet, are rewarded for guessing the same labels for pictures. It is generally an effective way to generate good labels for pictures (von Ahn and Dabbish, 2004). von Ahn also invented Captcha, a mechanism for discriminating people from bots when accessing sensitive information on the Web, such as a site to complete a purchase (von Ahn et al., 2004). This was elaborated into *ReCaptcha* (von Ahn et al., 2008), which uses the Captcha logic to identify words that have been poorly scanned in OCR documents (see Figure 2.5). This has proven to be an effective way to clean up documents that have been OCRed in, for example, the Google Books project. *Project Gutenberg* has a similar goal: distributed proofreaders correct digital texts that have been automatically digitized with optical character recognizers (www.gutenberg.org). A number of projects have used Amazon's Mechanical Turk to pay participants tiny amounts of money for completing a tiny task. One of the more remarkable examples of this is Kittur and Kraut's (2011) Crowdforge project that has used Mechanical Turk to have crowds of individual contributors create sensible texts. The non-profit Souper Bowl of Caring provides resources to local organizations which in turn contribute their stories and ideas for creative events (www.souperbowl.org).

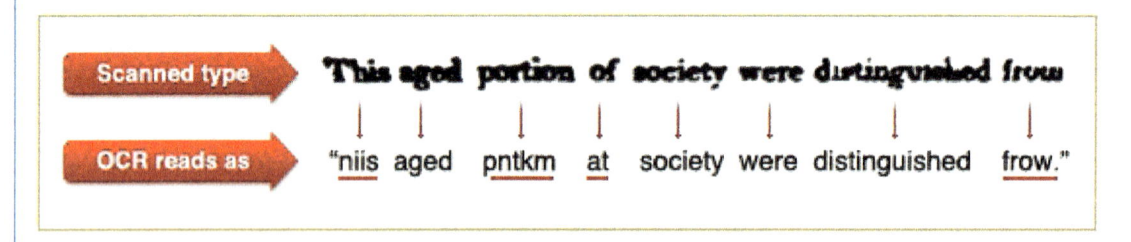

Figure 2.5: Example of the kind of material that would be used in reCAPTCHA to improve on scanned text. The items that OCR had gotten wrong would be presented to human users to identify.

The technical issues have to do with making the system reside on a variety of platforms and be easy to use. People have to be able to make their contributions without a lot of training, though some specialized cases like *Stardust at Home* have introduced a training routine that must be successfully passed before citizens are allowed to look for interstellar dust on images made during Stardust mission. Bonney et al. (2009) provide a good description of the process for developing a successful project along these lines.

2.5 VIRTUAL COMMUNITY OF PRACTICE

Virtual communities of practice are networks of people who share an area of interest and communicate about it online. They often share pointers to resources, best practices, upcoming events, etc. They are not focused on a shared project goal, like Distributed Projects. In science, from 2000–2008

Ocean.U.S. was an electronic meeting place for researchers interested in the oceans around the U.S., a resource for community building and planning, leading to the development of an Integrated Ocean Observing System (www.ioos.noaa.gov). The *Thesaurus Linguae Graecae* collection has a website that offers links, suggestions, job postings, etc., to researchers. WCENTER, a discussion list focusing on writing center issues (writingcenters.org), and *Tuesday Café*, is a real-time discussion group for writers (suzannesengl.blogspot.com; Sewell, 2004). Business schools often create executive forums or affiliate programs, e.g., for CIOs, where their advice and counsel are captured and put on a portal, and accessed by and added to by the members. They do not share information that would constitute a competitive advantage, but rather links and resources to make their professional jobs better.

Social networking software, such as Facebook, have provided a fruitful model for the creation of more specialized social networking sites for particular communities or organizations. One example is Dogear, a social networking site developed internally at IBM (Millen et al., 2006), which later was included in an internal environment called IBM Lotus Connections. Grudin and Poole (2010) looked at the use of wikis in enterprises, and found that while thousands of wikis had been set up within enterprises, only those that were for new groups in a start-up mode seemed to be successful in using this genre.

The main technical issue is accessibility and ease of use. Organizationally, the key issue is motivating participation and use. For more on the possibilities in motivating people, see Bos (2008) and Ling et al., (2005).

2.6 VIRTUAL LEARNING COMMUNITY

The purpose of a virtual learning community is to organize the high-level training around an esoteric topic. *ArchNet* is a collaboration that involves a digital library of over 600,000 images of Islamic architecture, and through active discussion boards participants learn about architecture while viewing images (archnet.org). The *Pan-Canadian Health Informatics Collaboratory* was developed by the British Columbia Institute of Technology as an e-learning environment to enhance existing or new education and training in the health sciences (Meginbir et al., 2004).

More generally, the world of on-line training and webinars (for web-based seminar) has grown strikingly in recent years. Such commercial resources as WebEx and GoToMeeting (also GoToWebinar) have provided convenient infrastructure for such activities. An expert on some topic can make a multimedia presentation, and often members of the audience can interact with the speaker either in real time or asynchronously. These tools are mostly available through any web browser, with the hosting taking place remotely. Audio can be carried either over the Internet (VoIP, or voice over Internet) or through a standard telephone line. These are more commonly used in corporate than in academic settings, since these commercial resources involve cost.

But increasingly, options are available for universities and other non-profits. An interesting, rich set of educational opportunities are emerging through the Khan Academy (www.khanacademy.org) and various Massively Open Online Courses (MOOCs) such as the OpenCourseWare at MIT and the new Coursera (Coursera.org). These offerings go beyond just broadcasting lectures, to include peer grading, which is necessary when courses enroll more than 100,000 students, which in itself is an active learning opportunity. And, they often foster the formation of local cohorts of students taking the same class, meeting face-to-face to discuss course material and exercises, as well as to foster a local network of people with common interests (DiSalvio, 2012).

Organizationally, as with any educational system, one wants to be sensitive to the diverse set of current knowledge of the students and meet their needs. Additionally, assessments of learning and credentialing have to be established to meet the needs of both the students and the professional certification board. An interesting example is a study by Gibbons et al. (1977) who showed that videotapes with a tutor can be more effective than standard classroom or even live video presentations. Cadiz et al. (2000) followed up on this earlier work by building cost-effective systems that allowed for this kind of learning, and again found positive results for supporting on-line discussion of video material.

2.7 COMMUNITY INFRASTRUCTURE PROJECT

A community infrastructure project's goal is to build the infrastructure or specific technical products so that people in a particular domain can build a collaboratory of various sorts. The Grid Physics Network (GriPhyN) joined information technology researchers and experimental physicists to build the first petabyte-scale computing environment for data-intensive science (Hofer et al., 2008). Digital Research Infrastructure for the Arts and Humanities (DARIAH) builds infrastructure to support collaborative projects in the humanities and related disciplines (www.dariah.eu). Both DARIAH and the Dutch Virtual Knowledge Studio (virtualknowledgestudio.nl) explicitly state that they want to open up the field to new ideas made possible by computation, data storage and retrieval, and networking.

Like the other kinds of collaborations, sharing of computational resources also requires developing standards, and managing large data sets. Issues arise in managing these as well. A study of the American Institute of Physics of their multi-institutional collaborations (AIP, 1992) found if these are managed by professionally trained managers, they come in on time and on budget; if they are managed by an academician, they are more likely to fit the sometimes quirky needs of academicians. In academics, a young person's career may be impacted if building infrastructure "counts" less than more scholarly publications.

2.8 REMOTE EXPERTISE

Many collaborations connect people with various kinds of expertise. Some collaboration focuses entirely on providing expertise to remote others. The most common such collaboration focuses on "telemedicine." An early effort, *EU-TeleInVivo* (Figure 2.6), concentrated on providing telemedicine mobile workstations with built in ultrasound devices in remote areas of the EU, for people on islands, other kinds of remote areas, and those in crisis situations.

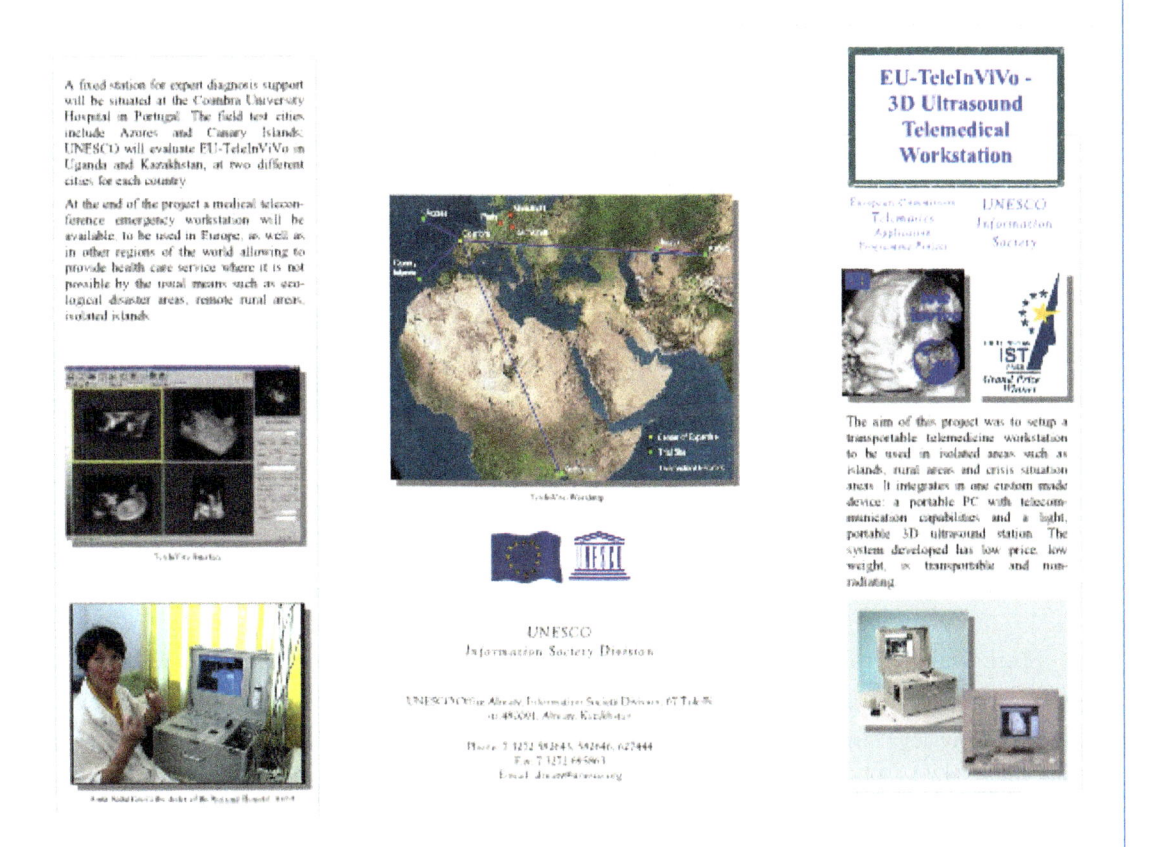

Figure 2.6: The EU-TeleInVivo project for remote use of ultrasound diagnostics.

More recently, InTouchHealth produces video-conferencing robots for remote medicine that can actually go on rounds and from patient to patient (www.intouchhealth.com) (see Figure 2.7). VGo, similarly, allows remote patient monitoring, as well as remote presence in school of children who are unable to attend for health reasons (vgocom.com). In the corporate world, there are occasions when large machinery breaks down, but the diagnostics and repair can be done remotely, mainly because much of today's machinery is computer controlled.

Figure 2.7: The InTouch Health telemedicine robot, with video conferencing on a remote controlled unit. Copyright © 2013 InTouch Technologies, Inc. Used with permission.

Technical issues focus on bandwidth and sensor technology (e.g., video, blood pressure readings) as well as security, since many of these projects involve health information. Organizationally, there are many legal issues about misdiagnosis and licensing across governmental units. Can an expert physician in Montana work remotely in Idaho? Additionally, the roles of the non-experts who are co-present with the device or person being diagnosed are likely to change. Do the nurses attending the remote robot physician have to now do more than they previously did, like open doors, administer diagnostic instruments and clear a path? But new business opportunities also open up for agents to represent the remote experts and be readily made available to the remote organizations that need them.

2.9 EVOLUTION FROM ONE TYPE TO ANOTHER

The world is not as simple as this categorization scheme suggests. Many collaborations start with a core of one type and then migrate to include other resources and services. For example, by using a shared instrument, one might find a colleague interested in the same phenomena and strike up a more focused collaboration. Many of the above-mentioned collaborations add information for the users that make it a virtual community of practice beyond a distributed project. As collaboration builds, often a community data system grows with it, along with the opportunity for outsiders to contribute and access the accrued data resource. Along with devising a shared instrument or modeling environment comes the need for a virtual learning community.

Common technical issues have to do with capacity to host other kinds of services. An early collaboratory, the *Upper Atmospheric Research Collaboratory*, a shared instrument collaboratory, started as a client application on NeXT machines which were purchased for the end users. By adding more instruments to the collaboratory and the emergence of the World Wide Web required a complete re-architecting of the system. Migrating from a shared instrument to a shared data repository added issues of server management and standards development (Olson et al., 2008).

Organizational issues also shift when one migrates to include a new capability. Often it has to do with how decisions are made; having a governance plan and an active community of stakeholders makes the transition go more smoothly. And, of course, migration also take significant funding; one is more able to make the case for additional budget if the core capability was successfully managed and delivered.

2.10 SOME OTHER FACTORS

Our typology focuses on the primary purpose of each collaboratory type. But there are some other factors that can influence how a collaboratory works, and what might be important to its success. We briefly list some of these here.

- **Size.** A collaboratory of a dozen scientists is a very different kind of activity than one of 3000 (like the CERN ones). Very different kinds of management practices emerge with scale.

- **Diversity.** A collaboratory in a single field is very different than those that are interdisciplinary (or multidisciplinary or transdisciplinary; see Stokols et al., 2013).

- **Multi-team.** An issue somewhat different than size is the extent to which a project consists of relatively loosely coupled teams. As defined in the introduction by the series editor, "Multiteam systems are two or more teams that interface directly and interdependently in response to environmental contingencies toward the accomplishment of collective goals (Zaccaro et al., 2013, p. xii)." This book reviews the complexities of these kinds of arrangements in great detail.

2.11 SUMMARY

Table 2.2 summarizes the technical and organizational issues highlighted in the paragraphs above. As we go through the success factors in the chapters that follow, it helps to know which issues are particularly salient for a particular type of collaboration.

Table 2.2: The technical and organizational issues particularly salient for each type of collaboration

Name	Technical Issues	Organizational Issues
Distributed Project	Rich media for communication	Stay motivated, develop standards, explicit inclusive decision making, intellectual property
Shared Instrument	Bandwidth, no delays, data transfer and storage, security	Access procedures, certification of users
Community Data System	Standardization, tools for modeling and visualization	Motivation to contribute
Open Community Contribution System	Usability, standardization	Quality control, motivation to contribute
Virtual Community of Practice	Wikis, accessibility and ease of use	Motivation to contribute
Virtual Learning Community	Technical disparities, interoperability,	Meet different educational levels, certification
Community Infrastructure Project	Meet the needs of the variety of users	Project management, impact on junior participants' career paths
Remote Expertise	Technical disparities, bandwidth, sensor technologies	Licensing across state borders, assistants' jobs, and new business opportunities.

2.12 KEY ATTRIBUTES

Table 2.3 shows how these eight types array themselves across two major dimensions: whether the collaborators interact with each other a great deal or not, called tight or loose coupling, and what is shared, tools, information, or knowledge (Bos et al., 2008).

Table 2.3: Relationships among the eight types of collaboratories

	Tools	Information	Knowledge
Loose coupling between collaborators	Shared Instruments	Community Data Bases Open Community Contribution System	Virtual Community of Practice Virtual Learning Community
Tight coupling between collaborators	Community Infrastructure Project		Distributed Project Remote Expertise

Shared Instruments and Community Infrastructure Project collaborations offer tools to others, but differ in how closely coupled the interactions are with each other. Community Data Bases and Open Community Contribution Systems collect their data or contributions, but again, differing on how closely coupled the interactions are among the collaborators. Where Community Data Base collaborators have to agree on standards, their contributions are quite independent after that. With the Open Community Contribution System, things that are contributed have to work together, but often the integration is guided by a central service or resource. Similarly, people in Virtual Communities of Practice and Learning share different kinds of knowledge yet are much more loosely connected than are people in Distributed Projects and Remote Expertise systems with the people they are consulting or collaborating with.

2.13 USING THE TYPOLOGY

We generated this typology to help see how collaborations differ and the technical and organizational issues that each type presented. After identifying over 700 collaborations that have distributed participants and revealing their solutions in technology and social practices, it helped us to categorize and find common issues and best practices. In subsequent chapters, as we go through the factors that affect the success of a collaboration, recognizing the type will inform the importance or value of a particular factor.

CHAPTER 3
What It Means To Be Successful

Before presenting the factors that both we and the literature confirm that lead to success in distance collaborations, we want to review what people mean when they use the word "success" in various discussions. It can vary from a new business or scientific breakthrough to merely that a newly developed tool was used by someone. We first review the criteria for success in the academic or research world, including both the sciences and the humanities. Then we turn to analogous criteria in the for-profit and non-profit worlds.

3.1 SUCCESS IN RESEARCH: THE SCIENCES AND THE HUMANITIES

Universities or large research centers are a special kind of non-profit, although many for-profit corporations have large research entities that also contribute in the ways we will describe here. Their goals are creating new knowledge, attracting talented faculty or researchers, providing hospitable environments for research and teaching, and passing that knowledge on to students or postdocs. Many scholarly disciplines, such as engineering, are interested in contributing to the solution of practical problems. Indeed, Donald Stokes (1997) argues that solving practical problems and contributing to basic understanding are not opposite ends of a continuum, but rather occupy the upper corner of a quadrant described by two dimensions corresponding to these two aspects of research. Thus, augmenting Cummings and Kiesler (2008), we will describe measures of success in five major categories:

- Ideas
- Tools
- Training
- Outreach
- Impact

The emergence of information and communication technologies has enabled geographically distributed collaborations that can bring together the right expertise, facilities, and resources to enable successful scholarly projects. Indeed, one reason why such difficult collaborations are on the increase, as described in Chapter 1, is that the successes thereby achieved can be spectacular.

But as we'll outline in this entire book, working together apart has many challenges standing in the way of success.

3.1.1 IDEAS

The most basic kind of success for scholarly projects is the creation of new knowledge. One of the longest running collaboratories in science, the Long Term Ecological Research Network (LTER), made a major discovery about the dynamics of hantavirus, a disease carried by deer mice but infecting only humans (Michener and Waide, 2008). Earlier we described how the Galaxy Zoo citizen science project allowed a Dutch schoolteacher to discover a new kind of galaxy. In general, such breakthroughs take a long time to create and recognize, and therefore it is important to measure other things thought to be correlated with ultimate breakthroughs.

The goal of many collaborations is to bring people from different disciplines together, counting on diversity to create breakthroughs. Stokols et al. (2008) argue that the emergence of widespread interest in the science of team science is often associated with bringing together ideas and methods from different disciplines. One possible measure of success might be the number of people actively involved and the number of disciplines successfully brought to bear on a problem. Cummings and Kiesler (2008) found, however, that though the number of disciplines involved in a project did not correlate with success (either positive or negative), the number of institutions involved did (negatively), suggesting this issue is more subtle. LTER's early and continuing goal was to bring people from other disciplines, like social scientists, statisticians, etc., to the study of ecology, and as mentioned, they have achieved some success. Many of the collaboratory projects that we have studied in the sciences and the humanities have united domain scholars with computer scientists to create tools that they hope will contribute to breakthroughs.

Some people measure the output of collaborations: higher quality publications, more patents, jointly authored papers, and any of these produced more quickly. LTER, for example, boasts over 18,000 papers as of 2013 and over 1,800 theses. The Cummings et al. (2012) longitudinal study of productivity in the NSF ITR program examined the number of publications, and specifically looked at net productivity controlling for a number of factors, including prior productivity of the individuals. Progress is also often thought tied to sharing ideas, so that people "can stand on the shoulders of giants."

Working more efficiently is also considered success, having fewer duplicated efforts from sharing ideas and results more quickly and more broadly. Google Scholar is one major contributor to fields whose publications are accessible from it.

3.1.2 TOOLS

Infrastructure projects (themselves collaborations) to support distance collaborations often explicitly report wanting to open up new ways of working. The BIRN collaboratories were test-beds to show how networking, shared tools, and coordinated federating of data could offer new ways of working (Olson et al., 2008). The Dutch Virtual Knowledge Studio for the Humanities and Social Sciences explicitly stated that they wanted to open up humanities and social scientists to new ways of working (virtualknowledgestudio.nl).

Capitalizing on broad contributions to a project, micro-participation, also represents a very new way of working. So-called "citizen science" has found ways to engage large numbers of people who do small tasks and then the results are aggregated (Bonney et al., 2009; Hand, 2010). Amazon's Mechanical Turk offers a mechanism to pay people small amounts of money to carry out a small task, and this has been used in a variety of creative ways in scientific research (Kittur et al., 2008; Paolacci et al., 2010; Kittur et al., 2011).

Associated with many collaborations is tool building specific to the domain. Many community data systems have tools allowing people to visualize and analyze the data. Success has many different levels in the area of tool building. That a tool gets built at all is one level; users adopting the tool (and complaining if it's taken away) is a second level. Tools moving from prototype to production quality is another level. New users adopting the tool is the fourth and highest level. As an example of the highest level adoption, the Electronic Laboratory Notebook created at EMSL has been adopted by a number of other collaborative projects. CORE 2000 offered a set of tools that have been built upon. *The Visible Human* added domain specific tools to CORE 2000 including image analysis (Myers, 2008).

3.1.3 TRAINING

All areas of scholarship require that new people enter the field and learn how to do research in one or more relevant areas. Thus, if a geographically distributed project is one that engages more entering scholars and facilitates their training, it achieves an important kind of success. Similarly, even established scholars need to update or expand upon their skills and knowledge, so training is relevant to all phases of a scholarly career.

Scholarly seminars can have broader reach through conferencing technologies like Centra Symposium,[4] Google Hangout, and Skype. We found these to be very effective for training in a multi-national HIV/AIDS collaboratory involving the US, Europe, and southern Africa (Bietz et al., 2008). We also observed a number of occasions where graduate students were mentored over the network by researchers at other sites (Olson et al., 2008).

---

[4] http://www.ivci.com/web_conferencing_centra_symposium.

New distance learning paradigms are emerging. One of the most remarkable is the new massive open online course, or MOOC. The core idea is making course material available on line, usually for free. Students can sign up, and may or may not be regularly evaluated, with some kind of documentation of completion at the end of the course. While there are many historical antecedents to the idea of offering large-scale courses over the Internet, these have really taken off recently. In the fall of 2011 Stanford University offered a course on artificial intelligence for which 120,000 people enrolled. A number of consortia have been created, including Udacity,[5] Coursera,[6] and edX.[7] Many major universities belong to one of these consortia, and the level of activity is accelerating. Many have found these to be remarkable opportunities to acquire general or specialized knowledge relevant to their area of research.

Having new ways of working with a broader set of idea generators can attract more and different kinds of people to the fields. Some of the beautiful images from the Upper Atmospheric Research Collaboratory (UARC) were available to schools, high schools and universities, and students could watch events unfold over time, both live and in replay, in the associated web portal, Windows to the Universe (www.windows2universe.org). Scientists used the replay feature to reanalyze and teach their colleagues. There was also evidence of the UARC/SPARC environment being used for remote graduate student mentoring (Olson et al., 2008). Schoolyard LTER reaches tens of thousands of K-12 students (schoolyard.lternet.edu). The infrastructure developed in the Great Lakes Regional Center for Aids Research (GLR CFAR) was used additionally for training health professionals about West Nile, new to the area (S. Teasley, 2000, personal communication).

3.1.4 OUTREACH.

Universities, like other non-profits, have to garner resources. Public outreach is key to persuading legislators and donors to fund academic endeavors. One measure of success is re-funding by federal funders or foundations. Just having the general public more interested (through outreach in schools and public presentations in the communities or in the media) convinces funders that there is a public good in supporting these endeavors. Another measure of success is having new funding sources appear.

Also, there are a number of early collaborations that inspired later ones. UARC inspired SPARC, and the BIRN test bed project successfully inspired other projects in the biomedical field. Many early participants in collaboratory projects became key figures in later projects, or in key administrative roles in funding agencies (see Olson et al., 2008), for examples pertaining to follow-ons to the UARC/SPARC decade-long project).

---

[5] www.udacity.com.
[6] www.coursera.org.
[7] www.edx.org.

3.1.5 IMPACT

As Stokes (1997) argued earlier, seeking deep understanding of phenomena does not have to be inconsistent with solving practical problems. His namesake example for this combination of traits is Louis Pasteur (hence the book title, *Pasteur's Quadrant*), who solved practical problems for French farmers at the same time as he achieved deep understanding of the germ transmission of disease. To quote Stokes:

> *"No one can doubt that Pasteur sought a fundamental understanding of the process of disease, and of the other microbiological processes he discovered, as he moved through the later studies of his remarkable career. But there is also no doubt that he sought this understanding to reach the applied goals of preventing spoilage of vinegar, beer, wine, and milk and of conquering flacherie in silkworms, anthrax in sheep and cattle, cholera in chickens, and rabies in animals and humans [p. 12]."*

Research can have impacts in many ways. We know that, as Kurt Lewin said, "There is nothing so practical as a good theory." A *theory* codifies complex relationships into manageable predictions. But in addition, research provides impact through *technical innovations* like new drugs, treatment plans, software, etc. Intermediate impact comes through new tools described above, like guidelines, patterns, toolkits and standards. Research can lead to *policy* that affects a number of people, like SOPA/PIPA[8] on intellectual property and privacy. *New media technologies* have made new ways of disseminating new knowledge available, including YouTube. For example, Hans Rosling made a number of video presentations that presented animated visualizations of population growth, education, and birthrates in different countries over the last century, engaging people in these phenomena. These videos have been viewed over 6 million times (www.gapminder.org). *Action Research* serves multiple goals: Gleaning generalizable findings while impacting people in a particular setting, like helping caretakers of pre-term infants monitor the infants' activity and other key factors (Hayes, 2011). And finally, as mentioned above, teaching to massive audiences as in Massively Online Open Courseware (MOOCs) as in Udacity and Coursera, open up new knowledge to very large numbers of people, not just students enrolled in educational programs but anyone anywhere.

One special case of impact is sustainability. For projects that last for longer periods, sustaining initial successes is important. Archambault and Grudin (2012) did a longitudinal study of the adoption and use of various social technologies at Microsoft, and while there was rapid adoption of many of them, some peaked and eventually declined. Of course, this is in a very complex organizational ecology, so sorting out exactly why these things happened is a complex matter.

---

[8] SOPA = Stop Online Piracy Act; PIPA = PROTECT IP Act.

3.2　SUCCESS IN CORPORATIONS

As we reviewed in Chapter 1, corporations are adopting what they often call "virtual teams" in increasing numbers. There are a number of reasons for this:

- needed expertise is not local, and employees are increasingly less likely to move locations (Lurey and Raisinghani, 2001);

- labor costs or taxes are lower elsewhere, increasing profitability;

- information about what global customers want and need is better gained by people local to them;

- not having to relocate may enhance job satisfaction, keeping human capital loyal to the corporation; and

- being able to learn new skills through remote learning opportunities or interacting with a remote expert increases value of an employee to the organization.

Corporations often measure success primarily by the size of their profits, which comes through increasing their productivity (output per cost of input and labor) or gaining better market share through higher quality or lower price. The recent Balanced Scorecard and Dynamic Multi-dimensional Performance (DMP) schemes for measuring success broadens beyond financial success and looks additionally at process and human capital factors that lead to financial success (Kaplan and Norton, 1996; Maltz et al., 2003).

The DMP scheme advocates that while there are some baseline measures in their categories, there are also some specific additional measures for type of corporation (large/small, new/established), and form specific measures that fit the specific domain of the company. Table 3.1 shows examples.

3.2. SUCCESS IN CORPORATIONS

Table 3.1: Examples of success measures from which to choose (adapted from Maltz et al., 2003)

	Baseline Measures	Firm Type Measures	Firm Specific Measures
Financial	Sales Revenue Growth	Stock price (large)	ROE Revenue per employee
Market/Customer	Customer retention	Market share (large)	On-time delivery
Process	Time to market	Depth of standardized processes (>3 years)	Quality initiatives
People Development	Retention of top employees		Employee satisfaction survey
Future	Anticipating unexpected changes	Investment in new technology development (>3 years)	Forecasting

Grudin (2004) provided a similar analysis in discussing how return on investment (ROI) should be thought about in the context of adapting new technologies. Table 3.2 shows his analysis, which draws on the work of McGrath (1984). He argued that too often the focus is on performance, yet this table shows that there are 11 other characteristics that could be important for the well-being of an organization. He offers as an example the introduction of changes in an administrative assistant's workplace that leads to a 10% performance gain, but also leads to 15% higher absenteeism and 20% greater turnover. Clearly, this modest performance gain would not be worth these other costs. In short, the question of success needs to be looked at broadly.

Table 3.2: Group functions (adapted from McGrath, 1984)

	Production	Group Well-Being	Member Support
Inception	Production demand and opportunity	Interaction demand and opportunity	Inclusion demand and opportunity
Problem-Solving	Technical problem-solving	Role network definition	Position and status attainments
Conflict Resolution	Policy resolution	Power and payoff distribution	Contribution and payoff distribution
Execution	(Performance)	Interaction	Participation

3.3 SUCCESS IN NON-PROFITS

Unlike corporations, non-profits are not out to make money; rather, their goals are to spend their money for the greatest good. Like corporations, efficiency and productivity is valued, because the same money can go further. Some professional organizations, like Charity Evaluator, will rate charities for how much money goes to management and how much to the people being served. They state that, for example, 7 out of 10 charities spend more than 75% of their money on programs and services; finding a charity that spends less is a red flag to inefficiency. Many other service non-profits rate things like the number of people served, number of food packages donated, and number of repeat visits by clients to services. Some evaluators measure impact in targeted concrete ways. For example, in a home-nursing non-profit for first-time mothers, they measured the birth weight of newborns and the number of babies vaccinated (Hicks et al., 2008). A non-profit whose goal was to reduce the number of illegitimate babies born and number of sexually transmitted diseases in the Bronx measured both direct outcomes and changes in attitudes through a detailed survey. But the impact, like a new discovery in science, the real goal, is hard to measure.

Moreover, success for a non-profit can be measured in terms of resources they attract. Perception of donors can affect their continuing to contribute. So, like corporations, brand name awareness is important. If funding comes from the government, political skills are important to secure continued funding. And, partnering with businesses, to get them to see the value in changing their practices, is a measure of success, both in financial and impact terms.

Finally, non-profits are also organizations, and the kinds of factors listed in Tables 3.1 and 3.2 are relevant here. A charity that burns out its volunteers with too much stress will not succeed in the long run. One of the evaluation criteria in a non-profit we studied included employee satisfaction, both with the management and with the efficiency of the service. Similarly, a public library, college, or social service agency faces the same issues in employee well-being and retention, as well as the quality of the service provided to their clients. Measures of success require a balanced view of these multiple criteria.

3.4 SUMMARY OF SUCCESSES

As outlined above, there is not single accepted measure of success. All organizations strive to use their resources well, to make progress in service, knowledge, or products. All organizations are more sustainable if their employees are equipped with what they need and have opportunities to increase their skill levels. Adopting tools built by others is akin to creating a product enjoyed in a market. Appreciation of the work by others, either customers or funders (shareholders, in some cases) creates continued wealth and resources to further the cause. To appreciate successes, we can point to a number of factors that are correlated with the ultimate success, be it breakthrough new knowledge, serving a large number of people well, or making a profit.

CHAPTER 4

Overview of Factors that Lead to Success

Given the increasing importance of distance work, we and a number of other researchers have attempted to both identify and categorize the factors that lead to success. We presented our initial categorization in "Distance Matters" (Olson and Olson, 2000), from which added various factors new to us in TORSC (Olson et al., 2008). Stokols et al. (2008) generated a list and summarization similar to ours, and Koslowski and Ilgen (2006) also have an organization that aligns well with what we present here.

We categorize the factors that we and others have found correlated with success as: The nature of the work, common ground, collaboration readiness, management and organization and technical readiness. We list these in the table below, phrased in the direction of what leads to positive success, and then explain each category and its details in the chapters that follow.

Table 4.1: Factors that lead to success in distance collaborations supported by various technologies
The Nature of the Work
Modular work is assigned to each location, requiring less communication overall
The work is routine and unambiguous
Common Ground
There was previous collaboration that was successful
Participants share a common vocabulary
If not, there is a dictionary
Participants are aware of the local context at other sites
Participants share a common working style, including management
Collaboration Readiness
Individuals tend toward extroversion, are trustworthy, have "social intelligence," and are, in general, good team members
The team has "collective intelligence," building on each other's strengths
The culture is naturally collaborative
The goals are aligned in each sub-community
Participants have motivation to work with each other that includes a mix of skills, the like working together, and there is something in it for everyone—not just a mandate from the funder.

34   4. OVERVIEW OF FACTORS THAT LEAD TO SUCCESS

	Participants trust each other to be reliable, produce with high quality and have their best interests at heart	
	Participants have a sense of self-efficacy, that they can succeed in spite of barriers.	
Management, Planning, and Decision Making		
	The project is organized in a hierarchical way, with roles and responsibilities clear	
	There is a critical mass at each location	
	There is a point-person at each location	
	The project manager:	
		Is respected
		Has real project management experience
		Exhibits strong leadership qualities
	A management plan is in place	
	A communication plan is in place	
	Decision making is free of favoritism	
	Decisions are based on fair and open criteria	
	Everyone has the opportunity to influence or challenge decisions	
	Cultural and time zone difference are handled fairly	
	No legal issues remain (e.g., IP)	
	No financial issues remain	
	A knowledge management system is in place	
Technology Readiness		
	The technologies fit the work	
		Communication tools have the richness and immediacy to fit the work
		Coordination tools (calendars, awareness, scheduling, workflow, etc.) are sufficient
		Everyone has access to shared repositories with sufficient access control
		Social computing (e.g., micro-contribution systems and social support) is well designed and fit the social as well as work needs
		Large-scale computation fits the needs
		Virtual worlds are used in appropriate ways
		The network has sufficient bandwidth and reliability
		The architecture fits the need for security and privacy
	The choice of technologies directly considers:	
		Speed, size, security, privacy, accessibility, richness, ease of use, context, cost, and compatibility

CHAPTER 5

The Nature of the Work

Distance collaborations are hard because today's technology, though increasingly sophisticated and decreasingly expensive, is a far cry from being sufficient. One needs technologies to support the myriad of things you get for free when you are collocated: awareness of the state of the other people, often knowing what they are working on, the environmental factors that may impact the work (e.g., impending tornadoes, blizzard), information about the person that engenders trust, etc. (Kiesler and Cummings, 2002). Very few people have the technologies in place to provide all that information. Distance makes **communication** of all you need to know difficult.

Social psychologists who study groups have long known that the nature of the task has a big impact on how the group works. As a result, they have developed a number of task taxonomies that have guided subsequent research on group behavior. Three of the better known are Thompson (1967), Steiner (1972), and McGrath (1984). McGrath's scheme focuses on the goals of the group work, and as such is not as helpful in determining roles and responsibilities being assigned to locations. Steiner, like our characterization above, was interested in how much coordinated activity was required by different tasks, and referred to this as the "task demand." This clearly is linked to the kind of coordination in collaboration that we know affects the ease of doing distributed work. A thorough understanding of Steiner's taxonomy can help with the design of work for distributed groups (see Table 5.1).

5. THE NATURE OF THE WORK

Table 5.1: Steiner's taxonomy of the nature of group work

Divisibility	Divisible	Subcomponents can be identified and assigned to specific members	Playing baseball Building a house
	Unitary	The task does not have sub-components	Pulling on a rope Solving a math problem
Quantity vs. quality	Maximizing	Quantity: The more produced, the better the performance	Generating lots of ideas Scoring the most goals
	Optimizing	Quality: A correct or optimal solution is needed	Solving a math problem Designing the best algorithm
Interdependence	Additive	Individual outputs are added together	Shoveling snow
	Compensatory	Decision is made by averaging together individual decisions	Citizen science projects Averaging ratings of job applicants
	Disjunctive	Group selects one solution or project from a pool of members' outputs	Having one art project represent the entire school
	Conjunctive	All group members must contribute to the product for it to be completed	Climbing a mountain Divide and conquer strategy
	Discretionary	Group decides how individual inputs relate to group product	Agreeing to vote on the best answer to a problem

Although the categories of "Quantity vs. quality" are less relevant to our goal here, the categories of "Divisibility" and "Interdependence" line up well with aspects of our "coupling" concept. *If* a task can be broken up, and many workplace tasks can be (it is divisible), then the issue is how dependent the people are of each other. The distinctions within Interdependence appear to focus mainly on how and who decides the form of the final output.

Thompson similarly delineates the interdependency of members in a group while doing a task, but highlights different aspects of the interdependence. His four levels are: *Pooled*, *Sequential*,

Reciprocal, and *Intense Interdependence*. *Pooled* interdependence exhibits the least dependence, similar to Steiner's Additive, above. Independent contributions are summed to produce the output, exemplified by a team stuffing fliers in conference totes. *Sequential* is a type of Conjunctive task, in Steiner's terms, where not only does everyone have to do their part, but also the output of one member must be completed before the next member can add their part. Processing travel reimbursements with a series of roles and approvals is a good example of the Sequential interdependence. In *Reciprocal* task interdependence, individuals rely on all the other participants to produce a unified whole at the end. Often the relationships between the parts and roles are quite complex, contributing to the difficulty. Conducting new product design is a good example of this type because it often involves complex interdependencies. In *Intense* Task interdependence is the most complex, because not only are people totally dependent on each other, their participation must be enacted simultaneously. Jury deliberation serves as this kind of interdependence because the exchange of information, persuasion, and coming to a final verdict continue with everyone present until they are done. The greater the interdependence, the more communication required among the team members.

Therefore, in considering stresses on distributed teams, it is important to assign work functions to people at different locations so that they don't have to have constant communication. **Modularization of the work** to the location is the goal (Chompalov et al., 2002; Birnholtz, 2006). We have seen a number of distance collaborations that did not modularize and witnessed the stress. If you are highly dependent on someone's work and do not communicate as if collocated, it is easy to drift and tempting to attribute to the other person faults in their personality or competence (Myers and Smith, 2011). Modularization was a design goal in NEESGrid, so that different components were developed in different locations and Application Programming Interfaces (APIs) negotiated in the requirements up front (Spencer et al., 2008).

People studying distance collaboration in the corporate world also now recommend reducing the **interdependencies** (Sonderegger, 2009). This flies in the face of one of the main reasons to have distant collaborations, to garner the contributions of experts no matter where they are (Gassman and von Zedtwitz, 1998). But the difficulty of being highly interdependent but not collocated negates some of the value of the expertise that could be available. IBM originally set up its financial report for products worldwide to be hierarchical by product, different locations being responsible for gathering the information for a product line worldwide and then aggregating the results at headquarters. This turned out to be so difficult that they switched after about 6 months to a scheme where each location was responsible for all products in its region, and the aggregation across regions done at headquarters. It turned out that resolving the different accounting practices in different regions was simpler when headquarters did it (once for all products) than everyone doing it repeatedly (once for each product in the region).

A second corporate example involved the joint development of an application to assess engineering designs for their manufacturability, an early assessment of which would narrow the design space and waste less design and testing time. Two locations, the U.S. and Mexico, were originally tasked to collaborate in the full design. This turned out to be extremely difficult, because independent designs moved forward without the communication required for tight coordination. Eventually, they reassigned the work so that the database design was assigned to those in Mexico and the software assessment engine to those in the U.S. This modularization required less constant communication, once requirements were agreed upon, and resulted in a successful final application. (Cummings et al., 2009) report similar examples.

EMSL collaborators purposefully divided their work into larger, more independent chunks over locations, and over time learned to communicate more, discussing analysis options and comparing notes at more intermediate stages (Myers, 2008).

A second factor in work that makes distance collaborations difficult is **ambiguity**. Herbsleb et al. (1994) noted that at Lucent, the process of building and changing the Class 5 Electronic Switching System (5ESS), the switch that at the time, half the U.S. telecommunication service providers used, is tightly coupled work and still succeeds. He explained that the development of the 5ESS had been going on since 1982, and the process is well known by all involved. The developers know who owns the code that speaks to their modules and vice versa. There is very little ambiguity in the process, and therefore, people know to whom to communicate when.

Unfortunately, ambiguity abounds in all kinds of work. Academic research is about things not known and often the researchers have to approach problems in new ways. Research and Development in corporations is similarly ambiguous. Even each new case of design has some measure of ambiguity. Consequently, special attention should be paid to ensuring high communication bandwidth whenever the work lacks an established routine.

CHAPTER 6

Common Ground

Two people have common ground when they share mutual knowledge, beliefs, and/or assumptions and know that they share it. It's not only the similarity but also the knowledge of what the other knows that makes conversations easy. Clark (1996) is the standard reference on the details of common ground and how it plays a role in conversation. Clark and Brennan (1991) explicated these ideas with respect to common communication media, and Monk (2009) has illustrated these ideas with respect to a series of computer-mediated communication (CMC) technologies. We will return to these latter points in Chapter 9 when we elaborate on the kinds of technologies that are used to support distance collaborations.

One clear indication of common ground is that people have worked successfully in the past. The experience alone gives them something in common and time to get to know what the other person does and does not know, and time to learn things that the other did know and one did not. The developers working on the 5ESS at Lucent had been working together for sometimes 20 years. Not only did they have similar training before working on the switch, the long-term experience was shared and they had time to know what everyone else knew. Often, when you have learned a corporate culture, e.g., AT&T, by being employed there, one can spot the embodied AT&T culture in people one hasn't met before; in fact their nickname was "Bell Heads," after the original company name, Bell Telephone (Steinberg, 1996). Cummings and Kiesler (2008) found that successful distributed collaborations often had participants who had had prior successful collaborations among themselves. Strikingly different culture has been the cause of unsuccessful mergers, like EDS and GM (Borghese and Borgese, 2002).

Many distance collaborations, however, purposefully mix people from different disciplines or backgrounds, in the hopes of generating innovation. In these cases, if people do not put extra effort into establishing common ground, difficulties will ensue. In a software development team at Andersen Consulting that we observed, it was clear after about an hour that there were different assumed interpretations of the word, "system," some meaning just the technology and others meaning the technology and its users and organization. One organizational response to this was to have "greenbeans," or new hires, assigned to most teams, and encourage them to ask questions when they didn't understand something. This often led to clarifications that were helpful to all, including the most experienced team members.[9]

Many collaborations have recognized the importance of establishing a common vocabulary. For example, in the BIRN project that focused on the mouse brain at all different levels of gran-

---

[9] Other aspects of these teams are discussed in Olson & Olson (1991) and Olson & Olson (2000).

ularity, the scientists discovered that they referred to the different parts of the brain with different words. Without agreeing on terms, they could not cross reference their findings in their databases. They developed both a Rosetta Stone translation scheme and further allowed reference in the user interface by pointing at a visualization of the area (Olson et al., 2008). Facing similar common ground issues, the scientists in the *Geosciences Network* (GEON) spent considerable effort in developing an ontology; *Lipid Metabolites and Pathways Strategy* (LIPID MAPS) had to develop a lipid classification system. Common ground goes beyond the words used; "When paleoecologists meet with contemporary ecologists, those whose work concerns deposits in the fossil record gesture and draw from bottom to top to signify the passage of time, while for contemporary ecologists, time flows left to right." (Hackett et al., 2008, p. 282.)

Distance without rich communication media also makes it difficult to share tacit knowledge. Informal conversations that are held around some referent object are often not recorded or conveyed to remote people. One solution at a large global automobile company, who had engineers in the U.S., France, and Germany, had rotators from each country serve in another country as the eyes and ears for their home team members (Olson and Olson, 2000). However, it's difficult or unwise to communicate politically sensitive information over email, because email is a record that is discoverable (Sonderegger, 2009).

When participants don't have a common location, they often miss out on different local practices that affect work. The large global automobile company with team members in the U.S., France and Germany often scheduled team meetings on what was Friday morning in the U.S. (headquarters), Friday afternoon in France. This regular meeting was scheduled during time that in France people normally did not work, having a 35-h rather than a 40-h workweek. Their participation was understandably "short," somewhat as it would be in the U.S. if people had to come on Saturdays.

On another occasion, with people meeting by videoconferencing from the London and Chicago, the Chicagoans were significantly late. Because those in London had no common ground about the weather conditions in Chicago, they were annoyed at the delay, to which they attributed lazy personalities, rather than the massive blizzard that was actually delaying the Chicagoans.

Not only may the knowledge differ with people from different areas of expertise, but also the working style. When work styles differ, people will generate less positive attributions of the others, feeling less empathetic (Cramton 2001; Williams et al., 2007). We found in UARC that domain scientists and computer scientists, building tools for the scientists, had very different working styles. The computer scientists had a loose way of deciding the requirements of the system, more like today's Agile than the Waterfall Method. The space physicists and the user interface designers expected a more planned approach with hierarchical management (Olson et al., 2008). Different management styles also plagued GEON. Sonderegger (2009) noted that international projects may falter because of training differences: research training in China and India is more hierarchical, more focused on implementation than on invention. Team members from the U.S. perceive them to be less independent.

Cross-cultural communication is fraught with differing expectations of conversational and collaboration behavior. We witnessed a three-way video conference between the U.S., France, and Germany, a regular status meeting but held on the day when one respected French engineer was retiring. The participants from the U.S., where "time is money," signed off the call as soon as the last agenda items were done. The Germans, who like other Europeans value and sustain relationships with more effort, stayed on the call and gave the engineer a humorous tribute, a "roast." The Germans and French were slightly taken aback at the Americans' failure to honor the French engineer.

Management styles also differ in different cultures. In particular, whether people give others negative feedback and how it is done differs remarkably between the U.S., Europe, and Asia. Browning (1994) has coined the differences in his "hamburger style of management:"

> "Managers start with the sweet talk—the top of the hamburger bun. Then the criticism is slipped in—the meat. Finally, some encouraging words—the bottom of the bun. With the Germans, all one gets is the meat. With the Japanese, all one gets is the buns; one has to smell the meat." (Browning, 1994)

On a more detailed level, conversational conventions differ in different countries. For example, in face-to-face conversations, people from the U.S. pause for about 2 s, waiting for someone else to take a turn. If nothing happens in 2 s they feel free to regain the floor. They fill up the silences. In Japan, the normal pause is 8–10 seconds, time to give in-depth consideration for what was just said (Pascale, 1978). So many conversations between U.S. and Japanese team members may be totally dominated by the U.S. members with them thinking the Japanese have nothing to say, when the Japanese team members consider the U.S. team members to be shallow and rude.

People from different cultures working together will adopt a common language, often English. This means that those for whom English is not their native language will be less fluid and misunderstand colloquial American expressions, like "home run," "baloney," and "beats me." It is important to have rich video connections when people who are not speaking their native language converse, both so gestures can help convey the meaning and for the speaker to see whether what was said was understood from the listener's facial expression (Veinott et al., 1999).

How decisions are made also differ remarkably in different cultures. In the U.S., we often vote with the alternative with the most votes winning (a democratic process) and the discussion has pros and cons in public discussion. In Japan, in contrast, decisions are made through consensus building often privately one-on-one; the decision then is announced in public. The American culture of being blunt and outspoken, without concern for the feelings of the idea generator, clashes with the culture of Japan where face saving is paramount (Pascale, 1978).

Cross-cultural collaborations can succeed, in spite of these difficulties. But team members have to be trained to both behave and interpret actions appropriately. This training often comes from living in the other country for a long time and having a helpful "translator" for not only words but also actions. In addition, there are online assessment and training tools, like Globesmart™ from Aperian Global Learning, that will allow team members from different countries to assess their values-profile and compare them with specific others on their team or with more generic data from those from particular countries. Where there are known differences, there are particular training modules on how to find a middle ground when confronted with mixed styles and values. Large organizations, like Google, have a corporate license for this tool and encourage those in cross-cultural teams to use both the personal assessment and training portions, although tricky cultural issues can come into play even when precautions are taken (Haines et al., 2013).

CHAPTER 7
Collaboration Readiness

Collaboration readiness has to do with individual characteristics that help people succeed in collaboration, what motivates people to work together, whether they trust each other, how well their goals are aligned, and how empowered they feel. In the corporate world, team members seldom get to choose with whom they work. However, there are still occasions where the people in different locations or in different departments are not fully cooperative, because of hidden incentives about who gets the prestigious work and who is valued when resources are distributed. Collaboration is even more complicated in the non-profit world, where typically the staff is employed, but they have to work with a myriad of volunteers whose motivations wax and wane, and are largely out of the staff's immediate control. In all these venues, the characteristics of the individuals and the collaboration climate can determine success or failure.

7.1 CHARACTERISTICS OF THE INDIVIDUAL TEAM MEMBERS

Some personalities work better in teams than others, regardless of distance or mixing disciplinary backgrounds (McCrae and Costa, 2008). Of the "Big Five" personality characteristics (extroversion, agreeableness, conscientiousness, emotional stability, and openness), extroversion is the one that most determines group behavior. Extroversion is a tendency for a person to gravitate toward other people. The others contribute to not only good work, but interpersonal trust among team members (Forsyth, 2010). Other characteristics that individuals have that make them good team members include self-management, good communication skills, cultural sensitivity (especially in cross-cultural teams), and comfort with technology and its changes (Blackburn et al., 2003). These are especially important in distributed teams because it is harder for people to monitor others, making the team member who is self directed and good at managing both their work and their behavior more likely to be effective (Duarte and Snyder, 1999). For those in transdisciplinary teams, having tolerance for uncertainties, flexibility, and being open to other perspectives turn out to be key (Stokols et al., 2008).

Two other individual characteristics have been shown to influence the effectiveness of a team are having good communication skills, and having members avoid dominating the conversations (Stokols et al., 2008; Woolley et al., 2010). In addition, there are individual characteristics that favor some individuals to being effective leaders in distance collaborations, which we'll cover in the chapter on Management.

Others have investigated the concept of "collective intelligence," the ability of a particular collection of people to outperform what would be expected both from their average intelligence or from the most intelligent in the group (Woolley et al., 2010). They found that the individual characteristic most influencing this collective intelligence is having social sensitivity, being able to read what's going on in another's thoughts through facial expressions and various actions. Often the more women in a group, the higher the collective intelligence, primarily because women have been found to have higher social sensitivity (Woolley et al., 2010).

7.2 THE CULTURE OF COLLABORATION OR COMPETITION

Although employees of corporations often do not have a choice with whom to work, they are not all naturally collaborative. Many consulting firms, for example, are internally competitive; people do not collaborate unless they are told to or unless they mutually need each others' skills, time, and expertise (Chatman and Jehn, 1994).

In academics, disciplines vary in their competitiveness. AIDs collaboratories are competitive because there is so much money and prestige associated with finding a cure. People in the Bio-Defense Center did not share their data because of fear of being "scooped," someone publishing findings from their data before the originator could (Teasley et al., 2008). A researcher in marine mammal behavior studies spoke of his data, "Well, honestly, I'm very protective about it…I guess it rather bugs me that I have to do the work, and everyone always asks me for a CD…it's our scientific study," (Meyer, 2009). In GEON, "Paleobotanists and metamorphic petrologists collect relative small data sets at particular geographic sites. The intense personal involvement with the research site and the data collection may lead to the unwillingness to contribute such data to a large anonymous repository." (Ribes and Bowker, 2008, p. 315).

In talking about U.K.-based eSocial Science, "The reality of individual competition over discovery claims, grants, promotion and space in top-ranking journals is far removed from the ideal of openness and sharing of data and other resources promoted by the e-science vision." (Halfpenny et al., 2009).

Collaboration is always going to be easier in a sharing, cooperative culture (Bos, 2008).

7.3 EXAMINATION OF THE EXPLICIT SOURCES OF MOTIVATION

Examining the motivation people have for working together is important. Often the collaboration puts together a set of people with different skills or expertise. The Great Lakes Center for AIDS Research (GLCFAR) was one such collaboration that put together people with complementary skills, especially between basic scientists and clinical researchers (Teasley et al., 2008). But these collaborations are successful only if there is something in it for everyone (Grudin, 1994).

7.3. EXAMINATION OF THE EXPLICIT SOURCES OF MOTIVATION

On a number of occasions, domain scientists joined with computer scientists to develop technology support for the domain. Unfortunately, the reward structure for academic computer scientists relies on publications of new ideas fleshed out in a prototype, when what the domain scientists need is production level code. Atkins and his blue ribbon committee about cyberinfrastructure said, "….balance must be struck between the concerns of technology developers (e.g., novelty and uniqueness) and the concerns of user communities (e.g., reliability and usability)." (Atkins et al., 2003) The UARC collaboratory suffered from this mismatch; the computer scientists wanted to build prototypes of systems to illustrate new solutions, not thoroughly tested production code. Often, human-computer interaction (HCI) researchers are asked to join in a collaboratory to both design the user interfaces and evaluate the outcomes, neither of which are the bases for publishable research. Recent Clinical Translational Science projects funded by the National Institutes of Health are required to have an evaluator on the team, and instead of hiring a social scientist whose goal is to publish, they often have hired a professional evaluator consultant to do the job.

GEON was fortunate to have had significant domain science and computer science problems to solve in the area of modeling continental growth (Ribes and Bowker, 2008), making it rewarding for all. Recognizing this problem of misalignment of goals, BIRN hired professional software developers instead of academic computer scientists on their teams to address this problem of alignment of goals. The software developers didn't need to find publishable results; they were motivated by their paychecks.

The motivator that we have found to fail over and over in science is when the funder mandates collaboration (Olson et al., 2008). Often the only collaboration that takes place is in writing the proposal. When the award is made, the people at the different institutions/locations divide the money and work independently. If the funder requires collaboration and the teams are not naturally collaborative, then to get collaboration one would have to write contracts stipulating it and the way it is going to be measured (Shrum et al., 2001).

As mentioned in Chapter 2, in Community Data Base systems, a number of reward structures need to be put in place to motivate people to share their data (Bos, 2008). *GenBank* requires genomic data to be entered into the database as a precondition for publishing. An especially interesting experiment was attempted by the *Alliance for Cellular Signaling*. They made an arrangement with *Nature*, a highly prestigious journal, to have their editorial machine vet the "Molecule Pages," the standard format for the output of hard work by the scientists, but unlike actual publications. *Nature* editors would then certify this review process when young professors came up for tenure with these kinds of publications. While the Molecule Pages have been a central and successful part of the project, the experiment with *Nature* did not work. Molecule pages were hardly ever cited, and citations played a large role in the evaluation of scientists doing the research.

Social connections also operate as motivators. The community of scientists in Zebrafish Information Network (ZFIN) have common academic roots (Olson and Luo, 2007). Many of the key people who started ZFIN were trained by George Streisinger, who first studied zebrafish. Some collaborations explicitly engage in group identity building, including having bowls of M&Ms at each location, "Science at the Bar" social activities, signature drinks, among other things (Hackett et al., 2008). Arguments of persuasion often invoke altruism and the common good.

7.4 TRUST

A key ingredient in collaboration readiness is being trustworthy and trusting in your teammates. Greater trust, cohesion, and communication are correlated with increased productivity (Stipelman et al., 2010). People often trust people they have things in common with (Feld, 1982; Kraut et al., 1990) and engender trust by adhering to established norms (Bradner and Mark 2008) which, of course, can be different at different locations. And, when team members not visible to others, there's a lot more they can get away with. Dispersed teams are known to often have lower trust and greater interpersonal conflict (Jarvenpaa and Leidner, 1999; Hinds and Bailey, 2003; Hinds and Mortensen, 2005). If trust is missing, people have to spend time setting up explicit milestones and contracts about the timeliness and quality (Shrum et al., 2001).

There are three different general kinds of trust (Rousseau, Sitkin et al., 1998):

1. team mates will keep their promises, called "confident expectations;"

2. they will produce outputs with high quality; and

3. team mates will keep each other's best interests in mind.

In distance collaborations, at the professional rather than persona level, "it's hard to know people's qualifications" (Sonderegger, 2009), especially if the team members are from different cultures. For example, in a high-energy physics collaboration between the U.S. and Japan, Global Accelerator Network (GAN), we observed that instrument operators in the U.S. were required to have a Ph.D. in physics, whereas in Japan they were not. Each party assumed the other was trained the way they were, which turned out to be wrong. Trust was lost. We have also seen occasions when in remote meetings, the key up-to-date budget documents were not distributed to all the locations, yet decisions were being made about allocation of resources. It took years for the remote locations to trust the core site again and only after explicit trust-building activities and acts to restore trust.

When collocated, trust is built by sharing personal histories, such as one's background, training, and observed competence (Jirotka et al., 2005) as well as some social things that lead to finding things in common, the basis of trust (Bos et al., 2002). When collaborating at a long distance, less can be observed, and there are fewer social exchanges, leading to lower measured trust (Herbsleb et al. 2000). Trust is more easily built with video- and audio-conferencing to connect long distance teammates than text chat and email (Bos et al., 2002). People are more comfortable working at a distance if they have met face-to-face (Rocco, 1998; Sonderegger, 2009), but even using chat for social topics can help (Zheng et al., 2002).

Trust is developed differently in different cultures. In Asia, for example, there is trust for people in one's in-group, but not for people outside, whereas in the U.S. trust is assumed until evidence shows someone not to be trustworthy. In Asia, trust is not assumed and must be earned by evidence or recommendation from someone in one's in-group (Cai and Hung, 2005; Yuki et al., 2005).

7.5 GROUP SELF-EFFICACY

Teams that feel empowered are more likely to succeed than one that does not, a concept called "collective self efficacy" (Carroll et al., 2005), a concept very like that of "team emotional and social intelligence" (Hughes and Terrell, 2012). Like personal self-efficacy (Bandura, 1977), the belief that you can overcome obstacles, collective self efficacy is the belief that the team can overcome obstacles, like a shortage of funding or unforeseen events. Such teams are more likely to find work-arounds or additional resources rather than give up. A detailed look at this concept can be found in Hall et al., (2008) including pointers to an assessment tool for new or prospective partners in team science.

CHAPTER 8
Organization and Management

The previous chapters covered the nature of the work and the characteristics of the individuals who are collaborating, that it helps if they have common ground and are motivated to collaborate. This chapter focuses mainly on the things having to do with the manager and what the manager can do to ensure that the collaboration is successful. How the collaboration is organized and managed is very important to its success (Cummings and Kiesler, 2005; Cummings and Kiesler, 2007; Kirkman et al., 2012).

8.1 THE PROJECT ORGANIZATION

Although much of today's answer to flexibility is a flatter organizational hierarchy, it turns out this works well only for collocated teams, where it is easier to communicate and share context and tacit information. For distributed teams, work goes more smoothly with at least some hierarchical authority and designated roles and responsibilities (Hinds and McGrath, 2006).

Some projects have people who are distributed unevenly, so that not all locations have the same number of people. Some are "hub and spoke," with a headquarters or central mass with collaborators in many remote locations, each alone. Unfortunately, the fewer the people the more likely the site is considered less powerful. Small sites also evoke the "out of sight, out of mind" problem, where the isolates are unattended to, and their needs and progress invisible to others (Koehne et al., 2012).

Power and attention is more evenly distributed if each location has a critical mass of people. The more even distribution of people, however, presents its own challenges. O'Leary and Mortensen (2010) found that when there is critical mass at a number of locations, the individuals have a tendency to form "in-groups" and "out-groups," with a tendency to disfavor and even disparage the out-groups.

As mentioned earlier, multi-team organizations add additional complexity to the managing of a project. Zaccaro et al. (2013) review these issues in detail.

8.2 THE PROJECT MANAGER

Having someone with project management experience, someone who leads the planning and manages day-to-day activities, who knows how to manage risk, who intervenes in crises, who represents the team vision to outsiders, etc., is critical to collaborations. This is well known in the corporate world (Kirkman et al., 2012). In many scientific collaborations, however, researchers would rather allocate money to the conduct of the science rather than project management. But after a number of failures, project management experience is being seen as more valuable in academic research. For example, the Biomedical Informatics Research Network (BIRN) had each of its original four major collaborations headed by a project manager in addition to the principal scientist. The ultimate authorities in decision making were the principal investigators, but the project managers ran the day-to-day operations. NIH's Catalyzing Team Science workshop recommended that postdoctoral fellows serve as project managers, and that these skills be valued in their career paths (Mazur and Boyko, 1981).

Unfortunately in the academic world, people are often involved in a large number of projects simultaneously. Because federal grant money is increasingly difficult to get, people have been applying to more and more programs simultaneously. As a consequence, some principal investigators end up with more projects than they can manage effectively. In these cases, someone has to be delegated the authority to manage the activities or else the effort is doomed. Time and attention becomes the most valuable, and scarce resource. Self-help books and seminars abound about time management. More recently, people are recognizing that it's not the time that needs to be managed, but the precursor, how people decide to commit to doing something. And, the fact that work includes activities that are local and those that are remote presents an additional challenge. Research has shown that when people have two equally important tasks to perform, one with people local and one with people remote, the local one gets the time and attention (Bos et al., 2004; Fussell et al., 2004).

In the corporate world, new project proposals are often accompanied by a project plan which includes how stakeholders will be communicated with, roles and responsibilities of team members, the estimation and then monitoring of time and cost and quality assessments, as well as risk analysis. In the academic world, management of a project is often much more casual, and often therefore flawed. Project management skills are just beginning to be recognized as important for success. The larger the collaboration, the more significant the management issue (Cummings and Kiesler, 2005). Acknowledging this, some funding agencies (NSF, NIH, and the former NIGMS "glue grants") ask for management plans and roles and responsibilities in large grant proposals.

8.3 WHAT'S SPECIAL ABOUT MANAGING DISTRIBUTED WORK?

"Leadership is the process of influencing others effectively, and the process of facilitating individual and collective efforts to accomplish a shared objective." (Weisband, 2008). Leadership is a challenge on most teams, but being distributed brings additional challenges. The biggest challenges are that it is difficult to monitor team members work and it is harder to develop and maintain trust (Tyran et al., 2003). "When working virtually, **distance amplifies dysfunction**" (Davis and Bryant, 2003). When leading across space and often time, people use technology to support awareness and communication. The leaner media, like the text of email, makes it harder to establish social presence: harder to convey attentiveness, warmth, and understanding, all the things that establish and maintain trust. Extra effort is required to do things that collocation gives us essentially for free (Kiesler and Cummings, 2002).

8.4 WHAT MANAGEMENT INCLUDES

Leadership consists of managing the tasks the team is to do and managing the relationships between the manager and the team and among the team members (Tyran et al., 2003). Leaders need cognitive skills, interpersonal and relationship skills, and, especially in the world of distributed work, tolerance for ambiguity (Bikson et al., 2008). Leaders build the team, establish the working culture, plan, and intervene when necessary.

8.4.1 PLANS

"Plans are nothing, planning is everything;" "Plans are of little importance, but planning is essential." These quotes from Dwight D. Eisenhower and Winston Churchill, respectively, reflect that with changing circumstances, any particular plan itself will not hold. But the fact that people have done the thinking to come up with a plan prepares them for the changes. Management and communication plans and the thinking that goes into them, therefore, are critical (Walther and Bunz, 2005).

The management plan outlines roles, responsibilities, and the timing of the various phases of the work. The communication plan records how people will communicate, when they will answer email, for example, and when standing meetings occur. The more explicit the rules for communication, the more likely things will go smoothly. For example, saying that email will be read and acknowledged within 5 h, not necessarily answered fully, but acknowledged, is an explicit rule that often aids communication. One of the hardest situations to comprehend and interpret in remote collaboration is the absence of a response (Kalman and Rafaeli, 2011).

Successful communication plans also include regular planned meetings often by video- or audio-conferencing, along with face-to-face meetings in which significant communication as well as trust building can occur. BIRN, like many other collaborations, holds annual "all hands" meetings at which many working groups make progress on shared issues (e.g., institutional review board coordination, forming standards for data collection and reporting). The more complex and interdependent the collaboration, the more complex and frequent need be the communication (Maznevski and Chudoba, 2000). In remote collaborations, it is especially difficult to convey tacit knowledge. "We know more than we can tell" (Polanyi, 1961). Recognizing this, the ATLAS project has emphasized the importance of time in the same location, the CERN facility in Geneva. Much happens over lunch. They reported "leading by persuasion and managing by coffee" (Birnholtz, 2008).

8.4.2 DECISION MAKING

When decision making is opaque or has the appearance of favoring some over others, collaborators lose trust, an essential ingredient in Collaboration Readiness, above. This is especially true in distributed teams where some collaborators are invisible to the decision makers. Decision making needs to be free of favoritism, and have fair and open criteria, something called "procedural justice" (Kurland and Egan, 1999; LeFasto and Larson, 2001; Larson et al., 2002). It need not be a democracy, but everyone needs to feel that his or her opinions are heard and considered.

This is the view from the US. In other countries, however, the process of decision making unfolds very differently, the difference itself being a source of tension and misunderstanding. In India and Japan, for example, more people have to be involved in decision making, so "buy in" is acquired before the decision is made public, rather than after, as in the U.S. (Katz, 2006; Hooker, 2012).

8.4.3 MANAGING ACROSS TIME ZONES AND CULTURES

More and more collaborations involve people in distant time zones. Two issues create additional barriers in these situations: The further you go, the more likely you are to be faced with serious cultural differences, and the more time zones crossed, the less likely that team members can have a real-time conversation without inconveniencing at least one member (Tang et al., 2012). In the Sonderegger (2009) interview study of international teams, time zone differences was the most frequently cited problem. In working with people from India, those in the U.K. said, "We can only really schedule calls at 8:30 pm which is 8 am for them. So it's a bad time for us and a bad time for them." The rhythms of the days differ according to when the sites' workdays overlap. We experienced several times in Europe where they commented that "the U.S. just woke up." During this overlap time, there was increased vigilance of email and the possibility of phone calls.

People in different countries have different holidays, so one is working when the other is not. As a consequence, someone is unexpectedly unresponsive to email, and is therefore attributed to willful avoidance rather than simply that they are home on holiday. People in the U.S. find it difficult to understand all the Bank Holidays they find those in the U.K. have in May, in particular, where those in the U.K. don't fathom the U.S.'s Thanksgiving weekend (different from Canada's) nor the celebrations of Memorial Day, Labor Day, and especially Independence Day.

Cultural issues loom large as well. So many aspects of conversation, decision making, and even thinking are different when you cross cultural boundaries (Nisbett, 2003). Pause structures in conversation, who has permission to speak, how disagreement is conveyed, who must be included in a decision, who can speak with whom, are all different in different cultures. Interestingly, people interact with remote others differently depending on how far away they think they are, independent of how far they really are (Bradner and Mark, 2002).

Training can be effective in helping people understand what others mean and how to be better understood by those from other cultures. Aperian offers a number of training opportunities, and hosts an online resource called "GlobeSmart."[10] GlobeSmartTM includes an online assessment of the user's culture (assumptions, behaviors, interpretations of situations, and expected actions). It then encourages one's teammates to similarly be assessed for comparison, or allows comparison with generic values from other cultures, e.g., with a standard Chinese or Indian professional. It displays the differences on five dimensions, then leads the users through example video snippets of people who differ on that dimension having a conversation where misinterpretations abound. It follows with recommended steps to take to move towards a better understanding of the others' values and ways to find a middle ground.

Technology creates interesting barriers to smooth communication across cultures. When one travels and meets face-to-face, one accommodates to the home country; "When in Rome, do as the Romans." But often in technology-mediated communication like video-conferencing, people forget to accommodate, because they are talking from their own home culture. In video conferencing, where is "Rome?"

Some of the work of collaborating is finding the right people to consult with or coordinate with. Finding the people one has to coordinate with at other locations is easier if there is a designated point person at each location, the person who knows their collocated teammates well, and can direct inquiries to the right person. Some corporations go so far as to have people from remote locations rotate to the key locations so there is someone with loyalty to the remote team members who can convey to them key tacit or undocumented information, serving as their "eyes and ears" (Olson and Olson, 2000).

8.4.4 MANAGING LEGAL ISSUES

Team members' home institutions bring in other obstacles, often legal ones (Stokols et al., 2003; Stokols et al., 2005; Sonnenwald, 2007). Different institutions, like universities and companies, have different expectations about intellectual property, for example, each trying to maximize their own benefit from it. This issue became particularly fierce in collaboratories having to do with AIDS, with the prospect of huge payoff from discovered vaccines or cures. Collaborations that have succeeded have spent a great deal of time up front working out these issues; others progress for a while before running into this major disagreement and ultimately failing.

Institutional review boards, which control the safety of human subjects and the data they produce, interpret federal U.S. regulations differently. We have even found uneven cooperative agreements with other branches within the same university system. Institutions in different countries have different regulations affecting collaborations in the domain of biology, in particular in the transport of materials. China, for example, has strict restrictions on exporting blood samples for AIDS research (Luo and Olson, 2008). Some institutions have stricter rules about privacy, especially in medical research. eDiaMoND failed because they were forbidden to share x-ray images

[10] www.aperianglobal.com

for breast cancer screening all within the U.K. (Piper and Vaver, 2010). Similar restrictions apply in different countries to the transport of antiquities, the apportionment of liabilities, and policies for informed consent. These legal issues are well described by David and Spence (2010). In addition, there are ethical issues, for example, in putting data sources together that allow investigators to find things that are normally deemed private (Dutton and Piper, 2010).

8.4.5 MANAGING FINANCIAL ISSUES

When people from different institutions collaborate, there may arise some surprising financial barriers. Most often these issues concern sub-contracting and different modes of paying bills. This can be especially daunting in projects that cross country boundaries. We experienced such a barrier in a collaboratory that crossed the U.S. with the U.K. and South Africa. In South Africa, money has to be in an account before anything can be spent, whereas in the U.S. expenditures are made and then invoiced to the funder. This difference was a showstopper until the accountants at both institutions found a solution by applying for a loan in South Africa to put the money into the account, and as things were spent, the invoice was paid in the U.S., effectively paying off the loan.

8.4.6 MANAGING KNOWLEDGE

For some collaborations, the point of the work is to collect larger sets of data than would normally be possible in one location. Function BIRN, for example, seeks to widen the collection area of early and late onset patients with schizophrenia to participate in having their brains scanned while they perform various tasks (Olson et al.. 2008). The National Ecological Observatory Network (NEON) is installing ecological sensors across the continental U.S. to collect data for a planned 30-year period to better understand climate change, land use change, and invasive species on the nation's resources and biodiversity (www.neoninc.org). For Google and Facebook, data are their business. They are all very good at managing their knowledge. This in general is the phenomenon of "big data" that is getting so much attention (Mayer-Schoenberger and Cukier, 2013).

Other types of collaborations are more casual about knowledge management, and that casualness often gets them into trouble. People have traditionally used email attachments and sometimes shared folders (e.g., Sharepoint) to manage their collaborative knowledge. How documents, drafts, notes, meeting minutes are stored and accessed today is both increasingly easy and at the same time confusing to people who don't plan this aspect of their collaboration explicitly. Google Drive, DropBox, and Office 365 offer storage in the cloud, but because the sharing mechanisms differ, people make mistakes, deleting documents that others need and changing organizations that make it difficult for others to find things (Marshall and Tang. 2012; Voida et al.. 2013).

Additionally, collaborations that anticipate long tenure have to worry about digital archiving. The challenge is how to migrate both the data and the computational systems as technology changes. Who will be able to access President Obama's email in 50 years? The Protein Data Bank serves as a good example of successful migration: their work began in the era of punched cards, but now has migrated to servers to hold its results of crystal structural analysis (Berman et al., 2004).

8.4.7 LAUNCHING A DISTRIBUTED PROJECT

Many of the projects we have studied benefitted from an initial face-to-face meeting of most or all of the participants. This contributed to the development of trust, and allowed the participants to tackle some of the early tightly coupled work that most projects face at the beginning. Such face-to-face meetings can also be helpful if the project continues for multiple years. Annual "all hands meetings" are a useful fixture of ongoing distributed projects.

8.5 SUMMARY

We have often been asked which of the various factors that contribute to successful long distance work is the most important. While we have no quantitative measures (at least yet) of the relative weights of the various factors, it is clear to us that good management is certainly one of the most critical factors. Of course, a good manager or management plan takes into account the other factors as well. But managing at a distance is very different than managing a collocated project, and it is essential that it be done well.

CHAPTER 9

Collaboration Technologies and Their Use

Effective collaboration happens when the tools needed are available and used appropriately by the collaborators. In this chapter, we first review the **kinds of technologies** that have been effective in supporting distributed work, with different kinds of work benefiting from different constellations of technologies. Our framework follows closely that of Sarma et al. (2010), listing technologies as **communication**, **coordination**, and **information repositories**, adding significant aspects of the **computational environment** (see Table 9.1). Although we may mention specific technologies, the point is not to recommend a specific current technology, because they will quickly be replaced with newer versions. Rather, we wish to emphasize the types of technology that are useful and why. This is in part an attempt to explicate the potential components of technology readiness that can be a source of success for long distance collaboration. Then we follow it with an analytic scheme to guide people to choose the right constellation for the work they will be doing and two examples of how different types of teams choose different constellations of technologies.

Table 9.1: Classification of technologies to support distance work
Communication Tools
Email and Texting
Voice and Video Conferencing
Chatrooms, Forums, Blogs, and Wikis
Virtual Worlds
Coordination Tools
Shared Calendars
Awareness Tools
Meeting Support
Large Visual Displays
Workflow and Resource Scheduling
Information Repositories
Computational Infrastructure
System Architecture
The Network
Large Scale Computational Resources
Human Computation

9.1 KINDS OF COLLABORATION TECHNOLOGIES

9.1.1 COMMUNICATION TOOLS

Email and Texting

Email is ubiquitous. It has been characterized by many as the first successful collaboration technology (Sproull and Kiesler, 1991; Satzinger and Olfman, 1992; Grudin, 1994; Whittaker et al., 2005). One of the cornerstones of its success is that today it is device or application independent; and with attachments, it is a way to share almost anything the recipient can read. As happens with other technologies, people use it for things other than the original intent. People use it for managing time, reminding them of things to do, and keeping track of steps in a workflow (Mackay, 1989; Carley and Wendt, 1991; Whittaker and Sidner, 1996; Whittaker et al., 2005). Because it was not designed for these purposes, it doesn't support these tasks very well, although gmail's recent attempt to glean "tasks" from email is a step in this direction. People develop ad hoc workarounds (usually involving paper) to do things like keeping track of who has responded to a request sent to many people. Special applications, like Doodle, were developed to coordinate various inputs from

9.1. KINDS OF COLLABORATION TECHNOLOGIES

many people, e.g., in this case indications of when they are available to meet. eVite's invitations are tracked for responses, and initiators can send follow-up emails to those who have not yet responded.

Instant Messaging (IM), sharing primarily simple text messages with another person or even a group, has made significant inroads into organizations. In some cases it has replaced the use of email, phone, and even face-to-face (Muller et al., 2003; Cameron and Webster, 2005). There is evidence that it is sometimes used for complex work discussions, not just simple back and forth about mundane issues (Isaacs et al., 2002). It is also used effectively for quick questions, scheduling, organizing social interactions, and keeping in touch with others (Nardi et al., 2000). A number of visual interfaces to instant messaging have been explored. An early example was Chat Circles (Donath and Viegas, 2002), where the amount of messaging associated with individuals was shown graphically. Another example was Babble (Erickson et al., 2002). Figure 9.1 shows a screen shot of Babble in action. Individuals are represented by colored dots, and the recency of their participation in a messaging stream is shown by how close to the center of the circle their dot is. Babble evolved into an IBM system called Loops and later Lotus Connections. The advantage these kinds of interfaces have for IM is that they give quick summaries of aspects of a conversation that would otherwise require scrolling through the history of messages.

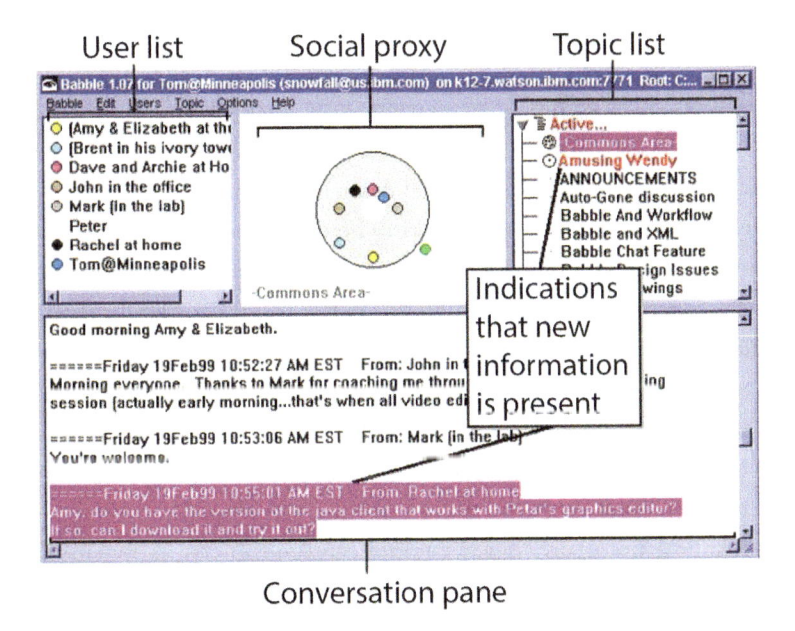

Figure 9.1: Screen shot of Babble. From Erickson, et al: Social translucence: Designing social infrastructures that make collective activity visible. Communications of the ACM - Supporting community and building social capital, Volume 45, Issue 4: pages 40-44. Copyright © 2002, Association for Computing Machinery, Inc. Reprinted by permission. DOI: 10.1145/505248.505270

60 9. COLLABORATION TECHNOLOGIES AND THEIR USE

Texting on mobile phones has emerged as a significant communication medium. It is much more common with younger users than with adults (see Figure 9.2). Texting can be a useful way to alert someone about an event, inquire if it's a good time to call, or share some small item of information. However, it—along with mobile phone conversations—has evolved into a major safety hazard when people do it while driving (Salvucci and Taatgen, 2011).

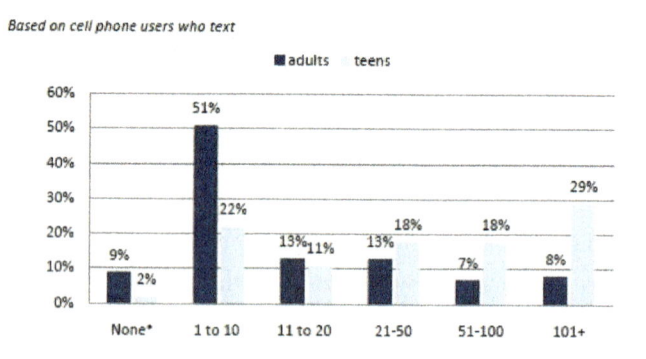

Based on cell phone users who text

■ adults   teens

Category	adults	teens
None*	9%	2%
1 to 10	51%	22%
11 to 20	13%	11%
21-50	13%	18%
51-100	7%	18%
101+	8%	29%

Source: Pew Research Center's Internet & American Life Project, April 29 - May 30, 2010 Tracking Survey. N=2,252 adults 18 and older; n=1,917 based on cell phone users. The teen results are based on data from June 26 - September 24, 2009 telephone survey, including cell phones, with n=800 teens ages 12-17 and a parent or guardian; n=625 for teen cell phone users.

Figure 9.2: Relative frequency of texting on mobile phones in 2010.

Mobile communication devices have social downsides, however. When jobs require employees to be available for consultation all their waking hours, the line between private life and work is blurred, often inducing stress and disconnects in one's private life. Some recent experiments in which employees are given "a night off" with others on the team covering for them, shows that not only did the employees enjoy the reduced stress and increased focus on the others in their private lives, but that by having to "cover" for someone, the team had a much better overall sense of what was going on (Mazmanian et al., 2005).

Except for the attachments (which can include elaborate drawings, figures and video clips), all of these so far are text based, and even thin text in the abbreviated world of texting. It has long been known that text does not easily convey emotion, and in fact can imply emotions that are not intended (Sproull and Kiesler, 1991). For example, conventional wisdom these days dictates that senders not use all capital letters unless they mean to shout. Emoticons emerged to help convey intended tone, such as a smiley face indicating a joke or kidding.[11] Of course, when the conversational partners know each other well and have a lot of common ground, these problems of interpretation

---

[11] A few examples of lists of emoticons: www.cool-smileys.com, messenger.yahoo.com/features/emoticons/.

of tone are mitigated. But text remains an impoverished medium compared to the tones and facial/body expressions possible in voice and video.

Conferencing tools: Voice and Video.

There are a myriad of opportunities to communicate beyond text in today's world, and many are used heavily. The telephone trumps text in being able to convey tone and to have immediacy of response. In fact, delays caused by technical interruptions of voice and video transmission are highly disruptive to conversational flow because of the importance of pauses in turn taking (e.g., Johnstone et al., 1995). With good connectivity such delays can be minimized, but it is still frequently the case that delays are present when connectivity is slow or erratic.

Many people have telephones these days that have the capability to easily do multi-party connections at least on a small scale. Organizations often provide services for larger scale audio "bridges" for conference calls. Key to the smooth execution of these calls is whether the phones have "full duplex" or "half duplex" transmissions. Half duplex lines are capable of transmitting only one direction at a time. Since natural conversations often include "backchannels," the "uh huh," "hmms," etc. that convey whether the recipient is agreeing, understanding, or not, when using a half duplex line, these are silenced. As a consequence, often the speaker will speak longer than necessary, not sure if the recipient has understood yet or not (Doherty-Sneedon et al., 1997). Additionally, conversational turn taking is often signaled by an utterance from the one who wants to take the turn while the current speaker is speaking (Duncan, 1972). These are entirely cut out in a half duplex line, creating awkward competitions for who will speak next.

Voice and video are far richer media than text. While tone of voice can add meaning to the words said, facial expressions and body language add another layer. In large meetings, video helps convey who is present without an explicit roll call,[12] and by eye contact and expression, conveys who is paying attention. But there is an emotional component as well: people will often say, "It is so good to hear your voice." Many report the extra "presence" someone has when they are connected by video (Bradner and Mark, 2001). One cannot only see the people, but the situation or context they are witnessing. For example, it is easier to understand the tone of a meeting where participants seem to be eager to wrap it up when an increasingly heavy snowstorm is visible out the window.

The richness of voice and video, however, can create barriers to people who are from different cultures. The expected pause structures in conversation are different in the Western and Eastern cultures, often creating miscues. Because Westerners are used to a shorter pause structure than Easterners, they will dominate the conversation (Ulijn and Li, 1995). Easterners, expecting a longer pause between utterances, pauses that convey respect, find the Westerners rude; Westerners interpret Easterners' silence as their having nothing to say. Similarly, when video shows facial expressions

---

[12] Although there can be problems if there are limited camera angles that don't reveal everyone who is present.

and eye contact information, because those features are interpreted differently in different cultures, people again may make wrong attributions of interest and consent.

As various services like Skype, Google Hangout, and Microsoft Lync[13] become readily available, more and more people opt for this rich channel. Skype has become the embodiment of the dreams of PicturePhone of the mid-20th century (Noll, 1992). High-end commercial video conferencing has achieved high definition and is increasingly better synced with the audio stream, and has good social ergonomics, but the installations remain expensive. Because the interfaces to these systems have not begun to reach the standardization of the telephone, they require experienced help in connecting, with the people often coordinating their connection through the more reliable connection of chat or phone.

Merely connecting with video is insufficient to achieve the "presence" of good connections. Eye contact and gaze awareness are key linguistic and social mediators of communication (Kendon, 1967; Argyle and Cook, 1976). In video, as in real life, people tend to focus on the face of the person they are talking with, and attempt to make eye contact by looking at the eyes of the person. Unfortunately, to appear to make eye contact requires the person to look not at the projected eyes of the remote person but at the camera. Therefore, to convey eye contact, extra effort needs to be expended to move the video of the remote person as close to the camera as possible; on many of today's big screen monitors with the camera on top, this translates to as high as possible. Without this careful adjustment, the camera conveys a sideways glance or the top of a person's head, which are interpreted as disinterest (Grayson and Monk, 2003). A number of experimental systems have been developed to provide eye contact (Acker and Levitt, 1987; Buxton, 1992; Ishii and Kobayashi, 1992; Okada et al., 1994; Vertegaal et al., 2003), but none of these are in widespread everyday use.

Another aspect of face-to-face communication that is being introduced to online tools is the ability to show objects and refer to them. The popular, simple way to do this is to send all participants slide decks and to direct the changing from one to the other by "next slide." But there are more sophisticated tools like GoToMeeting and Skype screen sharing that allow someone to share their desktop or a particular window with others, allowing them to control what others are looking at and being able to focus attention by using the mouse/pointer. In their study of distributed designers of an airplane part, Bradner and Mark (2008) found that a shared integrated object view called "3D-View" put individuals' drawings together, each with their author's name, so that they could see if there were conflicts and who was responsible for the fix.

A number of these communication channels are accessible on mobile devices. Their mobility has the advantages of being available from anywhere there is connectivity. But the downside of connecting in public places is the loss of privacy. Indeed, in an ongoing war zone like Iraq, being overheard in a conversation can be dangerous because the people nearby may be hostile. Many peo-

---

[13] A successor to Microsoft Communicator.

ple in such insecure environments resort to texting instead of voice connectivity explicitly because text cannot be "overheard" (Mark and Semaan, 2008).

Chatrooms, Blogs, Forums, and Wikis

Longer conversations from larger numbers of people are usually accomplished through chatrooms, blogs, forums and wikis. Although all have an expected response time longer than real-time conversations, the chatroom is often more immediate than the other three media. All can restrict who can participate to a designated work group or group of friends, but many of them are public. Often when the participants are designated, they converse using their own names; when it is public, the participants are free to use their own names or a pseudonym or nom de plume. Forums create discussions with entries and structured commentary, with anyone being able to start a conversation thread. In a blog, on the other hand, the focus is more on one person posting the material, with the expectation of and frequency of commentary much lower. This control difference creates different flavors of conversations, with the blogs covering topics that are in the blog owner's control and the forums often following the topical whims of the myriad of posters. Blogs with wide audiences often provide commentary on events of interest. For example, blogs played a major role in the 2004 U.S. Presidential campaign (Nardi et al., 2004; Adamic and Glance, 2005).

Wikis similarly are free-for-all conversations, but are even less structured in formatting. Forums are typically set up for discussion threads, whereas wikis can take any form whatsoever. Although their features afford typical structures, their features can be used in other creative ways. For example, wikis can be set up as structured forums or as shared documents similar to Google Docs. Grudin and Poole (2010) did a systematic study of the use of wikis in the workplace, and found that while management was often disappointed that they did not become generally useful repositories of corporate knowledge, they often were useful to support communication within and between teams. They found that wikis worked best with newly established groups, and with short-term activities.

Twitter is an example of what has come to be called a micro-blog, where the author posts contributions less than 140 characters, called "tweets." Others "follow" the tweets of individuals or organizations, being notified when the person/organization of interest adds something new. Tweets can be grouped by the use of hashtags, a "#" followed by a word or phrase. Twitter Lists can be used to create groups of tweeters that one would like to follow. It has become widely used, with roughly half a billion users in 2012. Various events have been organized, such as a Twitter Town Hall with Barack Obama in July 2011 that reportedly drew over 110,000 #AskObama tweets.[14]

Twitter has been used in a variety of ways. Jansen et al. (2009) looked at how tweets served as an electronic word of mouth, to share consumer opinions about various brands and products. Hu et al. (2012) looked at how Twitter was used to quickly disseminate news of current events, focusing on the death of Osama Bin Laden. Ross et al. (2010) studied the use of Twitter as a backchan-

---

[14] http://www.whitehouse.gov/photos-and-video/video/2011/07/08/impressions-white-house-twitter-townhall

nel during academic conferences, interestingly, focusing on humanists. Archambault and Grudin (2012) tracked the use of social media over a four-year period at Microsoft, and found that the use of Twitter grew rapidly at first, but then plateaued. But a series of studies shows that Twitter is common in the workplace (Zhao and Rosson, 2009; Ehrlich and Shami, 2010; Zhang et al., 2010).

An interesting article by Golbeck et al. (2010) looked at how members of the U.S. Congress were using Twitter. The did a content analysis of a sample of roughly 5000 tweets, and found that most of them (53%) were sharing information, such as points of view, and links to things like news sources, their own websites or blogs. Another 27% reported on their activities or their current location. A much smaller number (7%) were used to communicate with constituents. They looked at the stability of these proportions by taking smaller samples over time, and found the distributions were quite similar. They also looked at a small random sample of tweets by members of the U.K. Parliament, and found very similar distributions. Their conclusion was that Twitter was mostly being used by members of Congress for "self-promotion."

The content of tweets have been examined to figure out attitudes and beliefs. For example, Golbeck and Hansen (2011) looked at the political preferences of Twitter followers of a variety of news outlets. Marshall and Shipman (2011) used Mechanical Turk (see more below) to survey Twitter users about their attitudes toward a variety of actions regarding tweets, including archiving them, reusing them, and deleting them. Users reported frequently saving tweets of interest, but were more reluctant to republish them and still more reluctant to delete them.

Finally, Twitter has also been an important social medium in the rapid response by the public to disasters. In studies of what has come to be known as "crisis informatics," investigators have studied a wide range of natural disasters and the use of Twitter as an effective way to share timely information. Vieweg et al. (2010) looked at the use of Twitter in two natural disasters, the Oklahoma grass fires of April 2009 and the Red River floods of March and April 2009. Starbird and Palen (2011) looked at the use of Twitter following the 2010 Haiti earthquake. Huge amounts of useful information were shared in the ensuing tweets, and indeed these were often more up-to-date and accurate than the information in the hands of first responders. One possibility would be to develop data mining tools that could use tweets to extract information that would be valuable to first responders.

Facebook is a forum where, unlike Twitter, people don't just follow people they choose (like a celebrity), but have to establish a mutual agreement to be "friends." Facebook's trademark is to be yourself; in Twitter and other chatrooms, people often take the name and role of someone else. Sherry Turkle, author of *Life on the Screen* and *Alone Together*, reports being disturbed when she found a contributor to an early chatroom named "Dr. Sherry" (Turkle, 1995).

The fact that these discussion technologies are new has created confusion in some collaborative circles. For example, collaboration at IBM is supported by a toolkit called "Communities," which provides users (collaborators) with a set of tools they can drag into their designated collaborative space.[15] Whittaker (2011, personal communication) found that because the different features of the tools were not well understood, people spent an inordinate amount of time discussing the rules and expectations about how they were going to use them. Furthermore, when they were in use, again they spent a great deal of time arguing about things like whether a topic belonged in their forum or in the blog.

Finally, some of the public contribution systems have special features making them appropriate for particular purposes. At Google, every Friday the founders, Larry Page and Serge Brin, hold an open question and answer session called TGIF. To encourage the widest participation and use the time to answer the questions of most interest, they set up a system ("Dory") in which Googlers could submit questions to a pool and then many Googlers could vote thumbs up or down whether this was of interest to them personally. At the TGIF session, Larry and Serge would answer the questions voted most popular. This same system, renamed "Moderator," was then exported to the Obama campaign by a former Googler (Levy, 2011). The Obama campaign solicited open questions from people to answer during a public online forum. Ninety-thousand people submitted questions, and 3.6 million votes were cast on those questions. This system of open submission of questions and wide voting of which ones were of interest broadly is an important vehicle for an open democracy.

Virtual Worlds

Virtual worlds are graphical, 3-D representations of physical spaces, and have drawn considerable attention from both industry and academia (Bainbridge, 2007). They allow a person to experience a realistic environment, usually through an avatar. They can explore the space, manipulate objects, and, when networked together, can interact with other people's avatars. Such simulations of real worlds have been in common use for training in the military for a long time (Miller et al., 1995; Johnson and Valente, 2009). More recently, they have been widely available through such popular environments such as Second Life.[16] There is also now an open source virtual world platform called Open Simulator.[17] And multiplayer games such as World of Warcraft[18] have allowed for a wide range of playful interactions, though Brown and Thomas (2006) speculated that real leadership skills might be learned in a game like this that involves extensive quests involving substantial numbers of players. Virtual worlds have been used for a variety of real-world application domains, including health care (Boulos et al., 2007), scientific research (Djorgovski et al., 2010), software engineering (Koehne and Redmiles, 2012), and education (Wankel and Hinrichs, 2011). A particularly interesting case is what has come to be called "mixed reality," where the real and virtual worlds are combined in interesting ways (Ohta and Tamura, 1999; Herbsleb et al., 2000).

---

[15] This system emerged from an earlier one called Dogear, that got a lot of attention in the CSCW world (Millen et al., 2006).
[16] www.secondlife.com
[17] www.opensimulator.org
[18] us.battle.net/wow/

9.1.2 COORDINATION TOOLS

A class of technologies exists to support collaborators in finding a time to work synchronously, and a second set of technologies to support the coordination during the time together. Calendars and awareness mechanisms are in the first set; formal and informal meeting support tools fall in the second set. Workflow systems coordinate asynchronous work across different players, and complicated systems support the scheduling of shared high-end equipment.

Shared Calendars

Although the original introduction of group calendars was met with resistance, many organizations have seen value in their use (Grudin, 1994; Grudin and Palen, 1995). Calendars support the coordination of meetings, finding a time when the important participants are available. Resistance was fueled by those who hold an unfavorable view of the value of meetings themselves, not wanting to make the perceived time-wasting activity easier to schedule (Grudin, 1994). Others feared loss of privacy, with others making inferences about their performance by the kind and frequency of the meetings on their public calendars. However, with the advent of the feature of keeping the type of activity private (indicating only "busy"), and the ability to designate who can see and/or write on one's calendar, adoption has grown (Grudin and Palen, 1995; Mosier and Tammaro, 1997; Palen and Grudin, 2002).

Calendars are also used as a tool to display and/or read availability. When colleagues do not respond to requests in their usual timely way, one can view their calendar to discover that they are out of town or in a meeting. The information also allows one to plan when to contact a person (e.g., an "ambush" after a meeting in order to get a signature). And of course this can be particularly valuable for geographically dispersed colleagues, reminding people of where the workday overlaps and where not.

Awareness Tools

The topic of "awareness" has received a lot of attention in CSCW, and it is indeed a complex and rich topic (Schmidt, 2002). When people are collocated, there are a number of ways they learn whether others are available. In real life, open doors allow assessment of activity; noting that a colleague is not in the office but that their briefcase and coat are still there suggests the promise of return. Bumping into colleagues in the hall allows brief exchange of information. All of this is available without the explicit intention or actions of the people involved. When people are remote, these cues are missing and making cues available takes effort.

Early prototype awareness tools, such as Portholes, which gave colleagues a 5 s glance into a colleague's office, were resisted because of what was seen as an invasion of privacy (Dourish and Bly, 1992). More recently, awareness information is conveyed in the status indicators of Instant

9.1. KINDS OF COLLABORATION TECHNOLOGIES 67

Messaging (IM) systems. With IM, the user has control over what status indicator to convey to others, but it comes at the cost of remembering to set it and actually setting it. The cost of receiving the status setting, however, is very low. Many IM clients list the person's chosen buddies who agree to be monitored, and their status is typically listed in iconic form on the edge of the screen. Some researchers have built prototype systems that do not require explicit action on the part of the user, but rather send a message that is inferred from a number of sensor inputs, like the activity on the screen, the telephone off hook, the chair occupied or not, the door open or not, plus input of regular activities on the calendar (Gutwin and Greenberg, 1999; Horvitz and Apacible, 2003). Such awareness systems, however, are costly (in time and money) to install, so have not been widely adopted.

The systems described so far are those that indicate the user's current state, from which one can infer their interruptibility, but not often exactly what they are working on. In the domain of software engineering, where coordination of detailed efforts is of primary importance but the work nearly invisible, developers have created and widely adopted various system to "check out and check in" portions of the code they are working on. For example, Assembla[19] is a collection of tools to track open issues and who is working on them, plus a code repository where code is assigned to a person to work on, the time during which others are locked from editing. These kinds of coordination tools are powerful, but not widely adapted to domains other than software engineering.

A more general system that notes what people are working on in a shared document appears in Google Docs. The names of others who are currently editing the document are shown at the top of the document, their cursors with their names in a flag shown where they are working now, and their listing of past revisions, a listing of who did what (with authors' contributions highlighted in different colors) indicate what has been changed. These various indicators provide awareness of who is doing what, and who did what if more than one person is working on the document either at the same time or asynchronously.

Meeting Support

Coordination support for meetings, whether they are face-to-face or remote, comes in two flavors: formal and informal. The 1990s saw a lot of effort in Group Decision Support Systems (GDSS), where participants were led by a meeting facilitator through a number of computer-based activities to generate ideas, evaluate them in a variety of ways, do stakeholder analysis, prioritize alternatives, etc. (Nunamaker et al., 1991; Nunamaker et al., 1996/97). The clear wins in meetings supported in this way were that people could simultaneously generate their ideas and opinions anonymously without fear of negative evaluation, a technique good for eliciting more ideas. The downside, however, was that those with normal power and influence were unable to wield this power when things were anonymous. Although the decisions that arose from using these tools were measurably better, people did not like the experience (Kraemer and Pinsonneault, 1990; McLeod, 1992). Today,

[19] https://www.assembla.com/home

commercial products, like YnSyte, exist that embody these tools, and, although they are not widely adopted, they have been embraced by some organizations, like the military, where a large number of disparate stakeholders are involved and the algorithmic nature of the decision making process is naturally welcomed.

At the other end of the spectrum are informal meeting support tools, typically taking the form of a simple projected interactive medium, such as a Word outline or a Google Doc. The outline lists the agenda items as the highest level in the outline; during the meeting a scribe takes notes that everyone can view and implicitly vet. As agenda items are completed, the outline format allows the item to be collapsed, implicitly giving a visual sense of progress. Those applications that allow multiple people to author the shared document, like Google Apps, are even more powerful in these settings. When there is a single scribe, that person typically is so busy that they are barred from contributing to the conversation. When there are multiple authors "live," when one scribe talks, others can take over seamlessly to enter notes on what they are saying. Additionally, these note taking tools have been used very effectively in teams that have people for whom English is not their native language. The real-time visible note-taking is akin to "closed captioning" of the distributed meeting.

Large Visual Displays

A third type of coordination support is a combination of awareness (typically of a situation rather than people and their work) and rich information visualization. Large complex displays that allow monitoring of a complex device like a nuclear power plant or the nationwide communication network, serve both as shared information and sources of awareness (see Figure 9.3).

Figure 9.3: Large visual display for the Fermi Nuclear plant. Alexander Cohn, Times. File Photo.

A similar sort of display is used in large science. For example, scientists involved in monitoring the upper atmosphere, noting the trace of activity from sun storms and the concomitant distur-

9.1. KINDS OF COLLABORATION TECHNOLOGIES 69

bances in the upper atmosphere, first created a coordinated view of their sensing instruments. Over time, they co-developed an integrated rendition of what was actually happening, a view of the earth from above, side-by-side with what the theories said should be happening. This visual juxtaposition of theory and data allowed the rapid discovery of both conformance and dis-confirmations of data with theory, speeding the science (see Figure 9.4).

While the commercial world continues to produce larger and larger high-resolution displays at lower and lower costs, for some very specialized purposes wall-sized displays have been created by tiling together large numbers of displays. The Hiperwall, developed at the University of California, Irvine, and subsequently spun off as a commercial wall-size display, is a good example. See Figure 9.5 for an example.

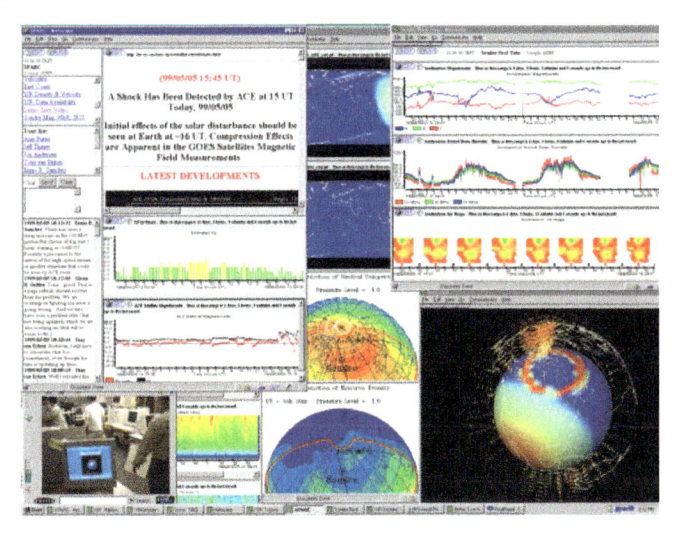

Figure 9.4: Visual display for coordinating data and theory in Upper Atmospheric science.

Figure 9.5: The Hiperwall.

Workflow and Resource Scheduling

A number of routine tasks that require input/approval from a number of people benefit from a structured digital workflow system. For example, the process of applying for a job involves a number of people: the candidate, the people who send recommendation support, the evaluators, and a manager who orchestrates the final decision, which is then communicated to the candidate, and when successful, the human resources department. A number of efficient online systems handle just this type of flow. A very successful workflow system supports the National Science Foundation grant submission, review, discussion, and decision-making process, notifying the appropriate players in the process at the appropriate time, giving them the tools and information they need, recording their actions, and sending the process on to the next in line. Although the rigidity of these system can sometimes prevent their adoption, a number of such systems have succeeded (Grinter, 2000).

In some research endeavors, especially in the hard sciences where the expense of a large piece of equipment necessitates researchers sharing it, systems have been put in place to schedule time on the equipment. Some clever systems have been created with the joint goals of being fair to those requesting time and to maximize the use of the equipment. Bidding mechanisms have been explored to optimize various aspects of the complicated allocation problem (Takeuchi et al., 2010). Various kinds of auctions have been tested to both create an equitable distribution of time and to prevent people from "gaming" the system (Chen and Sonmez, 2006).

9.1.3 INFORMATION REPOSITORIES

When people are collaborating, shared information needs to be organized and managed. The model of informally collaborating by sending people edited documents as attachments is common but fraught with challenges. All kinds of issues of version control and meshing of changes loom. A better solution is to have a place where the single document resides, a shared file, with all the authors having access. Many universities and large corporations provide such service as a matter of course; others do not. CTools, originally built as a learning management system headquartered at the University of Michigan, was found to be additionally useful as a project based shared file system. Administrators could assign read/write permissions to people for various documents and folders. In some cases, changes were made directly to the documents; in other cases, the collaborators made heavy use of version control, at the minimum through elaborate files names on copies.

Recognizing this difficulty, some large software solutions were generated. Microsoft, for example, offers Sharepoint, an integrated set of tools selected for use in a particular collaboration. It includes collections of websites, collaboration tools, information management (including tagging documents for permissions, types, and automatic content sorting). It also allows search over all the contents.

A more recent foray into shared editing and file management, but with a more fluid form, is Google Apps and Drive. The applications within Google Apps (documents, presentations, spreadsheets, forms, and drawings) each can be shared with others or placed into a folder, called a "collection," which can be shared. Each user of the document can put it in a folder of their choosing for their own view, allowing them to embody their own organization; they can additionally set whether the document, once put in a folder, disappears from view from the overview flat list. This myriad of features gives the users a lot of flexibility, but without vetted "best practices," many are doing it badly. And people are still confused by the "cloud" (Voida et al., 2013).

Those who share data rather than documents have an additional set of challenges. If data are being collected by a set of people, they have to agree, at the outset, what constitutes good quality data. Many scientific collaboratories have goals that include sharing data across sites. For example, in an early Biomedical Informatics Research Network (BIRN), they believed that progress on understanding schizophrenia would benefit from having a larger sample size of MRI images of patients, both with and without schizophrenia, doing various cognitive tasks while being imaged. A great deal of effort was spent in ensuring that the tasks that the patients performed were standardized, and that the various imaging machines were calibrated. In other scientific collaboratories, great care was given to developing a shared ontology of medical terms so that patient data could be aggregated from different locations and from different medical specialties, each of which had their own vocabularies (Olson et al., 2008). In the corporate world, global companies have similar problems in aggregating sales information because the accounting practices and codes differ from country to country. In one scientific collaboration, the scientists from different domains had very different names for the parts of the mouse brain, the model organism of interest. Instead of building a "Rosetta Stone" that translated one set of vocabulary items into another, they built an interface that represented the brain visually, allowing the scientist to merely point at the area of the brain of interest, without having to know any particular vocabulary (Olson et al., 2008).

In some areas of science and corporate research, key to the recording and vetting of information is the laboratory notebook. In many domains, the researcher keeps a personal record of their activities each day, including tests they ran, information they gathered, and things they noticed. It was important to sign and date each entry to record important discoveries, often feeding into patent applications. In some large scientific collaborations, they noted the value of being able to store and share these notebooks. They built an electronic notebook. One developed at the Pacific Northwest National Labs, the Electronic Laboratory Notebook (ELN) was so well designed that it was used heavily throughout the labs, and adopted by other collaboratories even in different domains (Myers, 2008).

An entirely different kind of information is also being stored for reuse. Some people find value in replaying certain events that unfold over time. Sometimes these are key meetings where decisions were made whose criteria need to be reviewed; in other cases, they are scientific events that may be understood differently under replay. And, increasingly, classes are being recorded for use by the current students studying for exams, and more widely in allowing students even from other universities or not even enrolled in any universities to hear lectures.

A more distant type of collaboration occurs when the people who are helping each other do not contact each other directly, but contribute their opinions and resources to a common repository. People ask for help on various sites where experts often contribute, like Stack Overflow[20] for programming questions. Early research prototypes, like AnswerGarden, attempted to answer people's frequently asked question on a topic sometimes without human intervention, once a "garden" of human generated answers was collected (Ackerman and Malone, 1990).

More recently, some services, like MovieLens and Amazon, use the large cache of people's preferences to recommend movies, books or products to others who have shown similar preferences. From experimenting with a variety of algorithms and presentation styles, research has found that recommendations are acted upon more often if explanations of why items were recommended are offered (Herlocker et al., 2000). Similarly, research showed that people are more likely to contribute ratings to such systems if they think their contributions are unique (Ren et al., 2007).

9.1.4 COMPUTATIONAL INFRASTRUCTURE

System Architecture

Many scientific collaborations have no choice as to how to architect their systems. The large-scale computation is either local or on a private grid of secure machines, and the data, often large, are stored on their own massive servers. Corporations similarly have architected their systems to be privately owned and operated, often sitting behind firewalls for security. There are challenges that organizations with firewalls face when attempting to interact with those outside the firewalls. While some kinds of Internet-based communication tools like Skype or Google Hangout can be used, more specialized tools like advanced video conferencing require special arrangements, such as working with a third party like Global Crossing[21] to bypass the firewall while keeping everything secure.

Today the general public has a choice whether to purchase applications for installation on their machines or to opt for computing and storage "in the cloud." If one chooses to work in the cloud, connectivity is important if collaborative access in real-time is required. Many cloud-based applications offer some level of off-line activity, though then the availability of up-to-date version control is lost. A more serious concern for some is security. We have encountered resistance to cloud computing from the medical world, military contractors, police and fire departments, certain

---

[20] stackoverflow.com
[21] In October 2011 Global Crossing was acquired by Level 3 Communications. More information is at www.globalcrossing.com.

government agencies, and others who are sensitive to information loss. While advocates of cloud computing often rebut such concerns, they nonetheless exist.

One interesting consequence of these different architectures is that each architectural choice creates its own behavioral consequences. When the applications and documents are on private machines, the mode of collaboration is hand-off, serial revision: documents are revised with "tracking changes" on and sent to the author-editor, who in turn can choose to accept each or not. The power resides in whoever the collective has made editor. In contrast, where the document and application resides "in the cloud," there is an implied place where those designated as editors can go to make changes. In this model, each edit appears as if accepted; the document is changed. Others can view the revision history and undo the changes, but at least at present a reversion to an earlier version undoes all changes, not just one at a time. Neither model in its current form is ideal.

These are two entirely different modes of collaborating in terms of workflow. Often, collaborators tacitly make the decision about who has the power to make changes, who can merely comment, and who has the final say in accepting the changes proposed. The existence of these two models present additional challenges to the users who are involved in collaborations of both kinds. They have to remember where something is stored, how to find it, and who has the power to decide on edits in each case, a situation we call "thunder in the cloud" (Voida et al., 2013).

The Network

Underlying all collaboration technologies is the network. Simply put, the bandwidth has to be sufficient for the kind of work to be done. Most of the developed world has adequate bandwidth for ordinary tasks, including video. Specialized needs that require large amounts of bandwidth will require specialized network infrastructure. Many of the large scientific projects have had to build high performance networks to handle the volume of data that comes from their instruments as well as specialized computing to garner enough resources to do the computation on that mass of data. For example, the ATLAS detector at CERN produces 23 petabytes[22] of raw data per second. This enormous data flow is reduced by a series of software routines that ends up storing about 100 megabytes of data per second, which yields about a petabyte of data each year. It requires special infrastructure to deal with data flows like this.

In developing countries, local telecommunication companies' facilities, policies and pricing can be quite limiting even for the basic level of communication. For example, technical support for an HIV/AIDS collaboratory that connected scientists from the U.K., U.S., and South Africa pushed for the use of videoconferencing, following the prescription for the richest media for communication across cultures (Bietz et al., 2008). However, the telecommunications infrastructure in South Africa at the time was inadequate for the requirements for videoconferencing or even for normal voice over IP (VoIP). Fortunately, a technology called Centra Symposium© was available

[22] A petabyte is 10^{15} bytes. 10^3 = kilobyte, 10^6 = megabyte, 10^9 = gigabyte, 10^{12} = terabyte.

that worked well with low bandwidth networks. Researchers were able to share visual materials such as Powerpoint presentations, and could use voice communication by means of technology that intelligently grouped the voice to take advantage of the natural pause structure of speech. As a result, the voice was intelligible, though the cost was a delay, on the order of 3-4 s between the U.S. and Europe and South Africa. A floor control mechanism allowed a speaker to hold the floor. Others could request the floor and a microphone icon could be clicked to pass the floor to another person. While it was a bit awkward, users soon acclimated to it, and successful voice conferencing over the Internet took place. There was also a text channel that allowed backchannel communication, the submission of questions, and troubleshooting. See Bietz et al. (2008) for details.

Network capacity is one of the more rapidly developing infrastructures. Indeed, the availability of high-speed wireless networks has enabled the enormous growth in mobile technology that plays a central role in today's long distance collaborations. It is possible to video conference from mobile phones and tablets, exchange photos and other rich media, and access material from databases anywhere.

Large-Scale Computational Resources

In many areas of endeavor, such as advanced scientific research or data mining in business, large-scale computational resources are needed. Certain high-end centers, such as the National Center for Atmospheric Research (NCAR), in Boulder, Colorado, have traditionally developed their advanced computational resources in-house. Similarly, computationally intensive businesses such as Google and Amazon have their own internal resources. But organizations such as the National Science Foundation (NSF), realizing that there is a need for advanced computing in many areas they serve, have built infrastructures to support advanced computation. The historically important supercomputer centers are one manifestation of this. Not only have high-end computational resources been developed, but associated high-speed networking to support access to these facilities have emerged as well. A particularly noteworthy example of advanced infrastructure to support such needs is the Grid, as sophisticated computational infrastructure that is widely used (Foster and Kesselman, 2004). A more recent example is the NanoHub,[23] a special computational infrastructure for nanoscience and nanotechnology.

Human Computation

There is also a tradition of using human capabilities aggregated over large numbers to achieve important computational outcomes. One of the earliest examples of socially organized human computation concerned the calculation of the return of Halley's comet, carried out by three people in 1757 (Grier, 2005). From the 18th century on there are numerous examples of collective human

---

[23] nanoHUB.org

computation in a variety of domains, as described in Grier's (2005) fascinating volume. These efforts ranged over many areas of scientific, mathematical, and statistical topics. Throughout the 1930s and 1940s many teams of men and women calculated tables of numbers of value to the military. Of course, after World War II many of these calculations were taken over by electronic computers.

However, this phenomenon has experienced a recent renaissance under the rubrics of crowd-sourcing (Howe, 2008; Doan et al., 2011), collective intelligence (Malone et al., 2010), the wisdom of crowds (Surowiecki, 2005), and citizen science (Bonney et al., 2009; Hand, 2010). The core idea is that in many interesting domains gathering together the small inputs of a large number of individuals ("micro tasks") can lead to results that can be as high in quality as judgments by experts and done in a fraction of the time. We covered a number of examples in Section 2.4.

A specialized resource that has emerged is offering micro-payments for little bits of work. Amazon's Mechanical Turk is probably the best known example of this, but it is hardly the only one. A variety of clever things have been done with Mechanical Turk. For example, Kittur et al. (2008) proposed using Mechanical Turk to do user studies, and provided some positive examples. Heer and Bostock (2010) used it to assess visualization designs.

9.2 DECIDING WHAT CONSTELLATION OF TECHNOLOGIES A PARTICULAR COLLABORATION NEEDS

As detailed in the previous section, collaborations typically need technologies to support **communication** and the sharing of the objects about which conversation takes place. Technologies are needed to **coordinate** the conversations, both to find times to converse and to coordinate around the objects. The objects, **information** and/or data, need to be managed: collected to exacting standards, managed, and made accessible.

Which technologies are chosen for a collaboration can have an impact on the success of the collaboration. In listing the technologies, above, we noted a number of features that affect how they are used, with choices at almost every juncture. In this section, we organize the discussion by features of the technology, highlighting the choices available that impact behavior.

A number of behavioral analyses of collaboration technologies have been offered in the literature. Olson and Olson (2000) listed a set of characteristics that themselves built on the scheme introduced by Clark and Brennan (1991), noting how well various technologies reproduce the "gold standard" of face-to-face interactions. Some time ago, Hollan and Stornetta (1992) pointed out that technology makes possible many things that cannot be done in the usual face-to-face encounter. For example, in face-to-face interactions, it is mostly impossible to be anonymous. As another example, in text-based systems, it is possible to review what you've written before sending it, whereas in real-time voice conversation it is not. What we offer here is a scheme that combines important features of technologies to support collaboration, whether the goal is to achieve the richness of real time face-to-face or to go beyond.

In short, the features that are important include:

- speed of response;
- size of the message/data or how much computation is required;
- security;
- privacy;
- accessibility;
- various kinds of control;
- richness of what is transmitted;
- ease of use;
- context information;
- cost; and
- compatibility with other things used.

9.2.1 SPEED

Conversation is directly affected by the speed of transmission. In face-to-face conversation, the utterances' speed and the structure of the pauses are well understood signals in the conversation flow. Bandwidth and other network characteristics can disrupt this flow, chopping up chunks of conversation and inserting artificial pauses. Since some networks are insufficient for video transmission, some system will similarly delay the speech stream to synch with the video. This is a mistake. It is much more important to have the speech stream convey real-time speed and pauses; video can either be delayed or ignored or turned off (Krauss et al., 1977).

Speed in conversation not only affects the initiators, in their expectations of a response, but also the receivers themselves in how quickly they can process the incoming information. Those for whom the spoken language is not their native language often prefer text over voice because the text can be read more slowly, understood more easily in spite of accents, and the expectation of an immediate, well-formed response is lower, allowing time for more careful composition. Therefore, faster is not always better.

Of course, other technologies generate different expectations as to how quickly the receiver will hear/read and respond. Voice, video, instant messaging, and texting are conducted at the speed of real-time conversations, with the exception of voice mail, which conveys expectations like email. Email, on the other hand, evokes a whole range of expectations of how quickly a response will happen. Because of this variability, we recommend that collaborators exchange information on their response habits and agree on a convention of promised responsiveness, key parts of what is called a "communication covenant."

Blogs, forums, and wikis set expectations for a much more leisurely pace of change. Some bloggers publish daily, others when the mood or events dictate. Forums, similarly, move at the pace of the trigger of key events and goals or deadlines. Wikis, being the most free form, can generate expectations and performatives of the genre its form imitates. Social computing applications reside at the very "slow" end of the response spectrum, with information for recommendations or votes aggregated over long periods of time.

9.2.2 SIZE

Modern messaging systems differ widely on the size of the contribution. Twitter is limited to 140 characters per entry; Facebook entries are larger, but not infinite; email can vary anywhere from short to very long, with attachments capable of conveying tomes. Documents, similarly, vary from paragraphs to tomes. Blogs and forums are typically no longer than popular magazine articles, with response to blog entries typically very short, where in forums they are a bit longer. One chooses the medium for text exchange at least partly on the basis of the size of each contribution.

Size also refers to the amount of computation required to process the entry. Text entry and editing require very little computation; spreadsheets and simple statistical packages require more, but still within the capacity of modern personal computers. Search algorithms have grown to require a great deal of computation in being able to not only find appropriate material but to rank it appropriately to the presumed needs of the user. At the high end of the spectrum for high-end computing are data mining, machine learning, and large-scale data analysis and computation, like climate modeling. Today's processors on personal machines can handle most personal and small-group collaborative computing needs, where as grid computing and huge server farms are required for the high-end computational problems.

9.2.3 SECURITY

Different domains differ widely in their requirements for security. Clearly, the military and military contractors require high levels of security as do medical establishments (because of HIPPA), the legal profession, and corporations with competitive secrets. To the people in these domains, a whole set of technology options are forbidden. Open source platforms and cloud computing are too insecure to be allowable for these activities. Simple, easy-to-use collaboration platforms like Skype and Google Docs are out of play.

Security also is a consideration at the individual level, where people in conflict situations (like a war zone or surrounded by competitors, for example, in a trade fair), need to use applications that cannot be heard or seen by "the enemy" (Semaan and Mark, 2012). Mobile phone user in these situations switch from voice to text; laptop users on tightly seated airplanes add polarized laptop screen shields to keep seatmates from reading what's on one's screen.

9.2.4 PRIVACY

In many applications used in a collaboration, it is important to know who contributed what. Often social conventions arise on shared documents that the editor ask the original author if they can make a change (Birnholtz and Ibara, 2012). In some cases, however, there is advantage to being anonymous or to holding a pseudonym (which over time and repeated contributions can gain a reputation, but just not an identity connected to a real person). Some public forums and chatrooms discuss controversial issues for which contributors could be in danger for their expressed opinions and therefore encourage anonymity. In addition, there are some group decision support systems that advocate that the participants be unidentifiable to encourage free participation without worries of reprisals (Nunamaker et al., 1991).

9.2.5 ACCESSIBILITY

By using the term accessibility here, we are not referring to systems that are usable by those who have some physical challenges, by being blind or deaf or immobile, though, of course, these are important considerations for the particulars of the participants in a collaboration. Here, we mean the more widespread issues of whether the use of a technology requires connectivity to the Internet or not. Those applications and storage facilities that are "in the cloud" are not accessible unless the user has access to the network and Internet. In contrast, applications that reside on one's computer, like Office, do not require connectivity to get work done, although they require connectivity to share.

In a similar vein, some items cannot be viewed or interacted with unless one has the appropriate application on their machine. For example, one cannot read and edit another person's Gantt chart unless one has the Project application installed. In other cases, the application or document is accessible to everyone, regardless of native applications. PDFs, today's email, and various web applications are more universally accessible.

9.2.6 CONTROL

Many collaboration technologies have features that automatically dictate who is in control or have settings to establish who can read/write, etc. Workflow applications designate at each step who can do what. The initiator establishes an entry or someone issues an invitation for an entry. At each of the next steps, it is clear who has the ability to add, edit, approve, etc. At the other end of the spectrum is an open collaborative document, like one in Google Docs or one in a file sharing system like Dropbox, where anyone can make any change. In between these conceptual ends is the implicit control over documents and their edits in a "tracking changes" situation, where one or more authors has the implied permission and authority to accept or reject changes, and to respond to comments. Although the explicit handoff and control in these applications works well for small well-defined collaborations, it creates coordination challenges when one author has to contend with the comments and suggested edits of a large number of colleagues, like in ad hoc committee work.

A similar set of control issues surround real-time meeting coordination where people are sharing an object relevant to a discussion. At one end of the spectrum are screen sharing programs, like GoToMeeting, where one person's screen is being shared, and that person controls the cursor and entry, but can explicitly hand control off to another. At the other end of the spectrum is the collaboration around views of a real-time shared editor like Google Docs. All participants can have the shared document open on their own screens, and can see the activity of all the other participants, both seeing who is "in" the document by reading a name on the upper right of the screen, and by seeing others' cursors with their name on a "flag" plus immediately seeing the changes as they happen. The downside is that not all participants may be looking at the same portion of the document, and conversations may be confusing when people say "here" or "there." Some prototype sharing system allow each person to have a private view of their entry and then can command when that view is either seen by others or copied to a public window (Monk, 2009).

This last point, control over others' ability for others to see one's work, has an interesting story to illustrate its dangers. To help coordinate the roll up of monthly sales figures around the world, a financial analyst sent around a sample spreadsheet to illustrate the format required. The spreadsheet had numbers in various fields that the immediate collaborators knew were mere examples, but more distant others who had access to the spreadsheet did not know this. Thinking these were real numbers, the others started talking about the implications and action steps, all to the chagrin of the originators and those in the know. As the creator of the example said, "Bad numbers have legs."

A similar concern over control, though at a more micro level, arises in the differences between various communication media. When conversing by voice, as soon as the utterance is made, it is received by the others. When conversing by text, however, one can review and revise one's message before sending it (Clark and Brennan, 1991). One has control over when the "final" version is viewable. This same point applies to the contrast between the immediacy of changes in Google Docs, and the more controlled release of changes in the handoff procedure in Office, using "tracking changes."

A more subtle level of control appears in the contrast between blogs and forums. In a blog, the control over the topic is held by the primary author; others merely comment. The others do not have the power to start a new topic or contribute large, complex material. In a forum, however, the control over topic is shared by all the members. Typically one or more people dominate, and even some take on the role of policemen to set and enforce the rules about topic choice. But the forum is much more democratic than a blog.

9.2.7 MEDIA RICHNESS

Many of the early prescriptions about what technology to use for what task (the "task-technology fit") centered on whether to use text, voice, or video (Goodhue and Thompson, 1995; Zigurs and Buckland, 1998; Fuller and Dennis, 2009). The early prescription was to use richer media for tasks that relied on the communication of emotion or subtle cues of intent. This pointed directly to tasks involving negotiation; one would like to see the opponent's reactions to various offers to assess the situation. However, the assessing another's reaction also plays out in communication between people who may or may not understand each other—those who have little common ground, or those who are speaking their non-native language to others similarly not speaking their non-native tongue (Veinott et al., 1999). Here again, rich media matter.

9.2.8 EASE OF USE

Some technologies are "walk up and use," and others require training and skill to use, often requiring the time of an operator or producer. The most prominent and varied of these collaboration technologies is video conferencing. An early functional but specially built system for video-conferencing, the Access Grid,[24] shown in Figure 9.6, is a very powerful, flexible multi-point system. However, it requires a "producer," not only to get everyone connected and audible, but also then to move and size the screens of the projected participants so that the person currently speaking is prominent, and others made smaller in a "Brady Bunch" peripheral configuration.

Figure 9.6: Access Grid video system from Sydney access point.

[24] http://www.accessgrid.org/

9.2. DECIDING WHAT CONSTELLATION OF TECHNOLOGIES 81

Commercial products like Cisco TelePresence[25] (shown in Figure 9.7) have achieved good social ergonomics by projecting people life-size as if around a table, but again it takes skill to know how to set the conference up and place the projected images in the right position. Desktop systems like LifeSize have a more user-friendly interface for connectivity, although it too requires training. The free and now very popular Skype video, audio and screen sharing program is easier to use, though less high definition and choppier in quality. Because it is fairly easy to use, it has grown in popularity, making it as close to the projected ubiquity of the PicturePhone as we have today.

Figure 9.7: Cisco's Telepresence with a social ergonomic arrangement.

The point about ease of use, however, is not restricted to videoconferencing systems. Not all information repositories are obvious how to use, especially in setting access and permission for actions. Some features of Google Docs are not obvious and best learned by watching someone more experienced using advanced features. Some special purpose applications for various scientific or high-end analytic situations, if not built with user-centered design principles, can fall into disuse because they are too hard to use.

Some collaborations, especially in technical fields, build their own technical systems. It is important in these cases that the systems be built with the user in mind, using Contextual Design (Beyer and Holtzblat, 1998), a key process for understanding the needs and capabilities of the users. The UARC/SPARC project, which built special tools for aggregating data and theory and

---

[25] http://www.cisco.com/web/solutions/collaboration/improving_collaboration.html

communicating about it for Upper Atmospheric physicists, had a user experience person on the team. The initial interface design matched that of the physical instruments in Greenland, the dials and meters familiar to all. Over time, the users themselves had direct input to the design of the follow-on iterations. That software migrated a number of times over ten years while the end users grew in sophistication about how digital data could be calculated and displayed.

Key to the success of introducing technology to people's work is what level of tools they have been using in their work already. In the UARC/SPARC example, above, at the beginning the scientists had familiarity with email, not interactive scientific models or complex integrated displays. With the increased sophistication of many of collaborators today, it is easier to have the discussion about what tools they need and what they are comfortable with, a concept we call Collaboration Technology Readiness.

Knowing that not all technologies and their collective use are perfect, it is important in the work to have adequate technical support, both for ongoing training, acquiring and adapting new tools as they appear, and maintaining the infrastructure. In the original BIRN project, a significant portion of the project funds went to their technical Coordinating Center, where they spared the scientists from having to decide what technologies to use by compatible suites of technologies for the scientists, what was affectionately called "BIRN in a box."

9.2.9 CONTEXT INFORMATION

Collaborators experience a number of challenges in awareness, not of each other's work or availability, but of what recent events have been or what the other is seeing at this moment in the conversation.

One kind of awareness turns out to be very important in establishing common ground in the moment (Monk, 2009). In face-to-face communication, we have co-reference; we know what the other is looking at; if I can see it he can see it. We can use deixis, using terms like "there" and "this" to refer to something we both know or can see. When conversing remotely with various communication tools, co-reference cannot be assumed. For example, if in a remote conference, both parties are looking at a copy of a large diagram, e.g., an architectural drawing, and one starts talking about a feature on the "upper left," confusion can often ensue. Unless the screens are locked together, what's on one's upper left may not be at the other's upper left. This same kind of loss of immediate common ground can occur in audio conferencing when the signal breaks up. One person says something they assume the other has received, but the choppy signal broke up at the key point.

A second kind of awareness is of the activity of others. Collaborative system may notify the collaborators when something has changed (e.g., a new document or a new version of a shared document exists), a "push," or the collaborator has to periodically examine the information repository for changes, a "pull." Email is the ultimate push; Facebook is the ultimate pull. "FACEBOOK is like a fridge. When you're bored, you keep opening and closing it every few minutes...to see if there's anything good in it" (J.D. Chandler, 2011[26]).

9.2.10 COST

Not all technologies are affordable by everyone. Top-end video conferencing suites such as in Figure 9.6 can only be afforded by organizations with large capital budgets. Those with fewer resources rely on inexpensive or even free options for doing video conferencing. Licenses for commercial software can range widely. And initial investments are not the only cost. Maintenance and upgrades are ongoing expenses. So in determining what hardware and software to acquire, these ongoing costs need to be taken into account as well. Developing one's own software with inexpensive student labor can have some short term gains, but may be very expensive in the long run if the software is unreliable or difficult to maintain.

9.2.11 COMPATIBILITY WITH OTHER THINGS USED

Not everything works with everything else. Hardware from various vendors can be incompatible, or at least be difficult to configure. Software is often incompatible with certain hardware. The worlds of Windows, Mac OS, Unix, Linux, and so forth can spawn a variety of interesting software options that only work within that world. Even the Internet, the World Wide Web, and the "cloud" can cause compatibility problems. And all of these matters can interact with technical support. Being a Mac aficionado in a Windows world can be a major headache when it comes to getting questions answered or bugs figured out. And of course there are elements of almost religious fervor in the selection of "worlds."

9.3 EXAMPLE DECISIONS ABOUT TECHNOLOGY CHOICES

Here we present two projects with different constellations of technologies to support their distributed work to illustrate the considerations of many of the factors listed above.

The first is a large-scale collaboration connecting 13 academic institutions in the U.S. with the goal of collecting standardized data and functional MRI images of both normal and schizophrenic patients to aid exploration into the diagnosis and cure of schizophrenia.

The collaboratory has developed their own suite of tools and methods for data sharing, query, and analysis tools. A major factor in their work is ensuring the consistency of the data collection situations and equipment, de-identifying the data, moving of large quantities of data between institutions, and building appropriate query tools to aid the research (Olson et al., 2008).

We contrast this with a team of 30-40 people doing user interface (dashboard) work in a large automobile company ("AutoCo"). The team has 25 people in headquarters, a few in Mexico

[26] https://twitter.com/jdchandlerklove/statuses/149628369703673856

and the rest in Germany. Individuals are matrixed, such that people with various expertise's work on several projects. People travel to meet with other team members, but also to vendors and partners to give them access to information that only AutoCo people can access. Security is paramount because they work on projects that can determine the competitive edge for the cars in a tough market.

Table 9.2: Comparison of an academic "large science" endeavor and a large automobile company about which technologies they use to support distance work

		Academic Collaboration	AutoCo
Communication Technologies			
	Email and Texting	Cross-platform, well understood. Lots of attachments	Microsoft Outlook, MSCommunicator
	Voice and Video Conferencing	A lot of conference calls. Video is available but people don't use it.	AutoCo internal conferencing service, Cisco's WebEx, sometimes Lync. Videoconferencing rooms available but this team does not use them
	Chatrooms, forums, blogs and wikis	Well-used wiki with posts of protocols, FAQs tips, and availability of new versions of software tools, articles of interest.	Chat in MSCommunicator including multi-way. No blogs or wikis because of security. Marketing using Facebook because they have to control what the public sees.
	Virtual Worlds	—	Global meetings done with Web Conferencing
Coordination Tools			
	Shared Calendars	—	Outlook Calendar
	Awareness Tools	Not needed because the focus is on shared data, not moment-by-moment communication	MSCommunicator, used frequently
	Meeting Support	GoToMeeting used a lot, with screen sharing	WebEx with screen sharing
	Large Visual Displays	—	—
	Workflow and Resource Scheduling	Dashboard indicating which patients have all their data in, and if not why not.	Cobbled together system to indicate who is currently responsible for an issue

Information Repositories			
	Shared Data	Clinical data is kept locally, then a search engine copies needed data to someone else's machine, "federated."	Sharepoint and Team Room, plus a very large Parts list
Computational Infrastructure			
	System Architecture	Tools are open source, data is federated at search time	All in-house, nothing in the cloud
	The Network	Open academic connectivity with safeguards in data transmission.	Two networks: Virtual Private Network and web facing applications
	Large-Scale Computational Resources	Earlier used Grid computing to merge with another collaboratory	Not in this team, but large simulations in design
	Human Computation	——	——

As shown in the Table 9.2, the selections the two contrasting teams made are very different on almost all choices. The Academic Collaboration is open source, using a variety of tools, many of which are built in house, with the exception of email and conferencing with GoToMeeting. The AutoCo in contrast has a tightly controlled set of commercial applications controlled by their central information technology department for the company as a whole, and all running on a secure network, with a few applications available through web interfaces. The work of AutoCo is fast moving with a lot of communication and coordination; the Academic Collaboration is slower moving, with a goal of coordinating large sets of standardized data so that many researchers can make progress in their particular research topics. They share findings and tips through a well-used wiki. They had solutions to almost all the categories of collaboration technologies, but because of circumstances of their work and the need for security, the choices were different.

9.4 CONCLUSIONS

Choosing the appropriate suite of technologies to support collaboration is not easy. The sets of features of each drive how they are each going to be used; the technology often dictates social configurations of use. Although we have not provided a decision tree of questions, answers of which would point to the "right" set of technologies, we have provided a listing of classes of collaboration technologies, a listing of the key features of these technologies that should be carefully considered in the choice of one's particular use, and two examples of how these played out in real distance collaborations.

CHAPTER 10

The Science of Collaboratories Database

Distributed collaborations to support scientific research have become very common in the past couple of decades. Starting with the early calls for "collaboratories" (Wulf, 1993), the community has embraced this concept and initiated numerous such projects in a wide variety of fields. We decided some time ago to track these developments and create a database of such projects. We now know that there are literally hundreds of them.

For more than a decade, we have been collecting some basic information about as many collaboratories as we can find. The goal has been to provide a public resource of collaborative efforts in academia, and to gain some insight into trends such as what academic areas are most active in what types of collaborations, which people are involved in a number of collaborative efforts, etc. These are primarily collaborative projects in science and engineering, but recently we have identified a number of emerging collaboratories in the humanities (Inman et al., 2004). As of August 2013, we have identified over 717 collaboratories. The primary means of data acquisition has been through snowball sampling. We have encountered collaboratories in articles, talks, and online mentions. While our sample is by now quite large, given the means of sampling we do not assert that our sample is representative.

10.1 INFORMATION COLLECTED

The information we have collected on each collaboratory includes:

- the name (e.g., Cell Migration Consortium (CMC));

- the url (e.g., http://www.cellmigration.org/);

- its status (e.g., operational, completed);

- its start and, if completed, end dates;

- the primary collaboratory function (e.g., shared instrument);

- the secondary collaboratory function (e.g., expert consultation);

- domain(s) (e.g., Biology, electron microscopy);

- a brief description of the collaboratory (often taken from their website);

- enumeration of the access to instruments they provide (if any) (e.g., electron microscope);

- enumeration of the access to information resources (e.g., shared repository of fMRIs of late onset schizophrenics);

- enumeration of the access to people as resources (e.g., consultation from the Beckman Institute staff);

- funding agency or sponsor (e.g., Beckman Institute for Advanced Science and Technology and the Illinois Consolidated Telephone Company);

- associated notes on the funding;

- organizations affiliated with the collaboratory (e.g., Norwegian Knowledge Centre for the Health Services) with the approximate number of people from each;

- total number of participants (approximately);

- communications technology used (e.g., ProCite5, email);

- technical capabilities (e.g., asynchronous object sharing: index/metadata, email attachments);

- key articles; and

- project-reported performance data (e.g., This project's web resources received over 50,000 hits on the website per quarter).

Figure 10.1 shows a typical entry.

10.1. INFORMATION COLLECTED   89

Figure 10.1: A representative entry in the SOC database.

Figure 10.2 shows the top level aggregate listing, which includes the name, a hot link to the website, the date of its beginning, and the primary collaboratory type. Each of the columns in the public view is clickable in order to sort on that variable.

10. THE SCIENCE OF COLLABORATORIES DATABASE

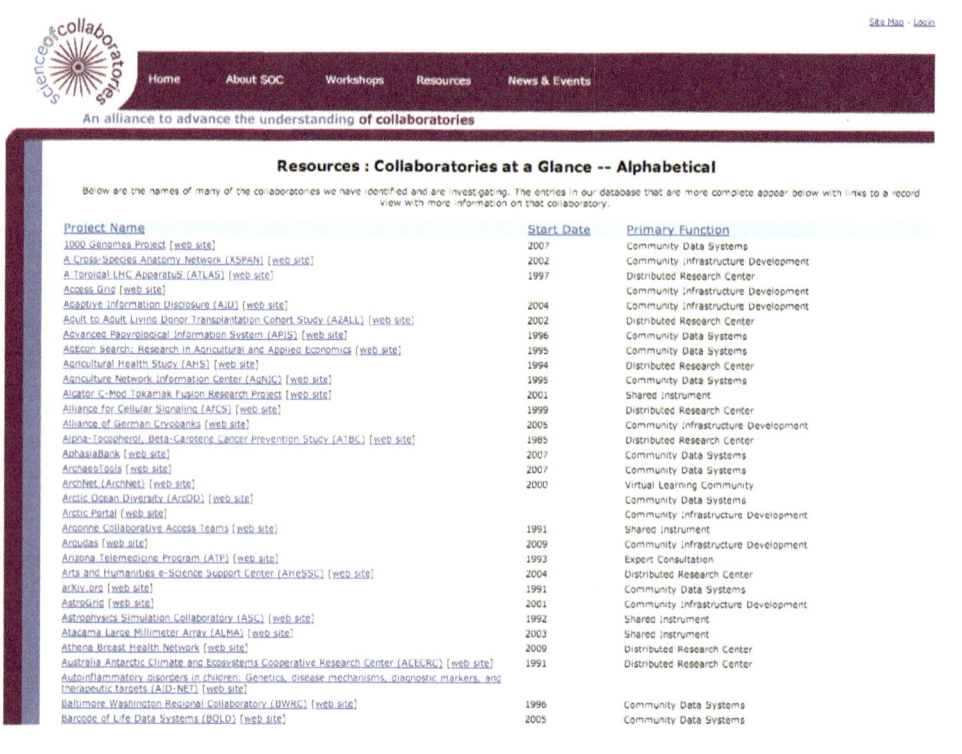

Figure 10.2: The top of the high-level listing of collaboratories.

SOC Database currently contains listings for 717 collaboratories, some more complete than others. The public view shows information at two levels: those collaboratories that have information in the above listed data fields filled in, and those that are merely noted with their websites linked. The database to date has almost 280 filled in, but more than 400 have little or nothing in their fields. The collaboratories that are filled in include (already) over 3,100 participating organizations and 67 different funders, and over 1,600 identified participants (with many more unidentified). The public view of this database is found at http://soc.ics.uci.edu/Resources/colisting.php.

A number of predefined reports are available to the research team (including those outside of our immediate research group we give permission to).[27] For example, Figure 10.3 shows part of a report of all the collaboratories that are associated with domain names in NSF format.

---

[27] If you are a researcher who would like to participate in the mining of these data, please contact Judith Olson at jsolson@uci.edu.

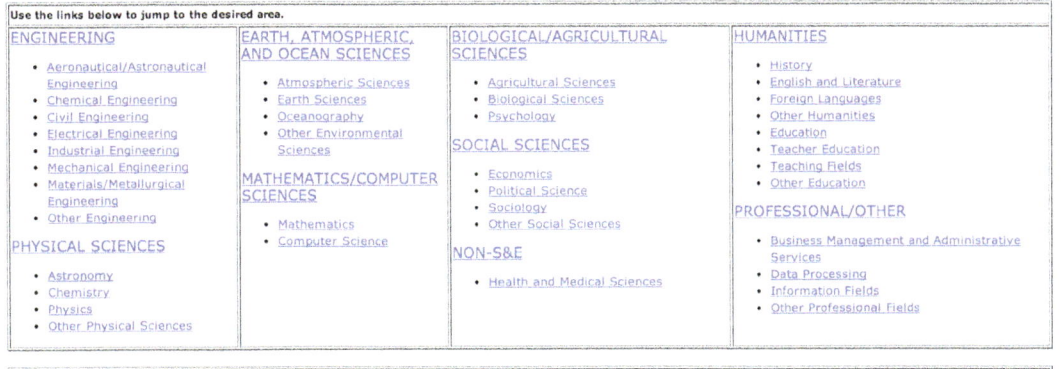

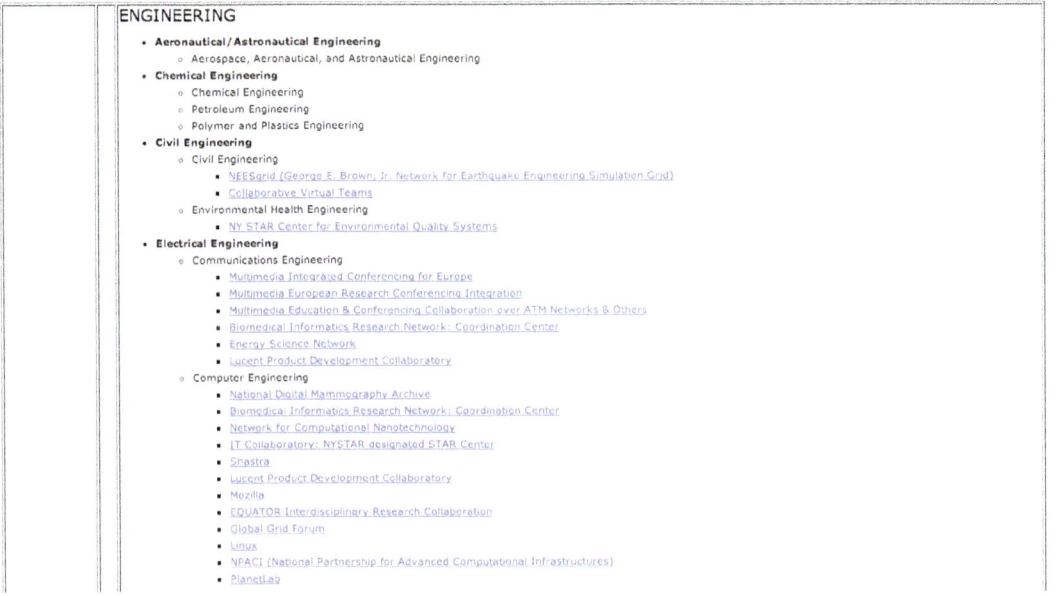

Figure 10.3: Example of the kind of report that is possible with the SOC database.

To the research team, a number of additional password protected items are accessible, including notes on how the collaboratory started, its relation to other collaboratories, a list of key people and publications, where the information came from (e.g., interviews, websites) and other investigator files such as interview notes.

10.2 FINDINGS TO DATE

In what follows, we describe some interesting summaries of the 278 Collaboratories whose entries we have filled in. Recall that ours is a snowball sample; we cannot generalize the entire population from our numbers. But the numbers are interesting nonetheless.

10. THE SCIENCE OF COLLABORATORIES DATABASE

All Collaboratories are tagged with the research domain name, using the National Science Foundation's classification system. Table 10.1 lists the number of collaboratories associated with each major category of domains. Note that one of the hallmarks of the Collaboratory movement is interdisciplinarity; consequently, these numbers include some collaboratories being counted more than once. In fact, the average number of domains per Collaboratory is 1.86.

Table 10.1: Major disciplines associated with Collaboratories		
Domain	#	Largest subcategory
Biological, Agricultural Sciences[27]	220	Biological Sciences
Earth, Atmospheric, and Ocean Sciences	82	Oceanography
Humanities	68	Education, in particular Science Education
Mathematics and Computer Science	59	Computer Science
Physical Sciences	52	Physics
Health and medical sciences	42	Health Sciences in general
Engineering	40	Electrical Engineering
Social sciences	34	Psychology
Professional	22	Information

Chapter 2 lists the types of Collaboratories, each having different foci, structure, and often participation. Table 10.2 lists the distribution of Collaboratory types among our 278 Collaboratories.

Table 10.2: The distribution of Collaboratories' major types	
Name	Number
Distributed Project or Enterprise	86
Shared Instrument or Resource	25
Community Data Bases	67
Open Community Contribution System	13
Virtual Community of Practice	17
Virtual Learning Community	20
Community Infrastructure Project	44
Remote Expertise	6

A number of institutions are very active in a number of Collaboratories. Table 10.3 shows the top 11. UCSD has been very active in the Collaboratory movement primarily, but not exclusively, through the Supercomputer Center.

---

[28] The NSF includes psychology as a biological science. We consider it a social science, and therefore have shifted the numbers associated with psychology to the following category.

10.2. FINDINGS TO DATE

Table 10.3: Institutions with a large number of Collaboratories associated with them

Institution	Number
University of California San Diego	35
Lawrence Berkeley National Laboratory	17
Stanford University	16
Massachusetts Institute of Technology	15
Harvard University	15
California Institute of Technology	15
Argonne National Laboratory	15
University of Illinois at Urbana-Champagne	14
Johns Hopkins University	14
Northwestern University	14
University of Washington	13

There are 58 funders in our database to date, with a number from Germany, the U.K., Canada, the EU, the Netherlands, Norway, Sweden, France, and South Africa. Table 10.4 lists the funding agencies that have funded the most Collaboratories.

Table 10.4: Funders of the largest number of Collaboratories

Funding Agency	Country	Number
National Science Foundation	U.S.	53
National Institute of Health	U.S.	44
National Institute of General Medical Sciences	U.S.	19
National Cancer Institute	U.S.	16
Federal Ministry of Education and Research	Germany	13
Department of Energy	U.S.	13
Joint Information Systems Committee (JISC)	U.K.	10
National Library of Medicine	U.S.	8
National Human Genome Research Institute	U.S.	6
Engineering and Physical Sciences Research Council	U.K.	6

We believe it is generally held that the collaboratory movement was generated by a need for "big science," primarily in physics, biology, and medicine. Although those are the domains of the largest number of identified Collaboratories in our database, the early Collaboratories represent a surprisingly wide set of types and domains. Table 10.5 lists the earliest Collaboratories we have in our database.

Table 10.5: Early Collaboratories and their domains

Collaboratory	Start date	Type	Domain
Inter-University Consortium for Political and Social Research (ICPSR)	1962	Community Data Base	Social Sciences
Protein Data Bank (PDB)	1971	Community Data Base	Molecular Biology
Energy Science Network (ESnet)	1974	Community Infrastructure Project	Communications Engineering
Ocean.U.S. (formerly ScienceNet)	1979	Virtual Community of Practice	Oceanography
GenBank	1982	Community Data Base	Molecular Biology, Genetics
Center for Economic Policy Research (CEPR)	1983	Distributed Project or Enterprise	Economics, Public Policy
Child Language Data Exchange System (CHILDES)	1984	Community Data Base	Developmental and Child Psychology
Alpha-Tocopherol, Beta-Carotene Cancer Prevention Study (ATBC)	1985	Distributed Project or Enterprise	Pathology
Carotene and Retinol Efficacy Trial (CARET)	1985	Community Data Base	Epidemiology
ITER, an international Fusion Collaboratory	1985	Distributed Project or Enterprise	Plasma and High-temperature Physics
Visible Human Project (VHP)	1986	Community Data Base	Biomedical Sciences, Anatomy
Centre for the Study of Learning and Performance (CSLP)	1988	Distributed Project or Enterprise	Educational Psychology, Measurement
Sloan Digital Sky Survey (SDSS)	1988	Community Data Base	Astronomy
Peirce Online Resource Testbed (PORT)	1989	Community Data Base	Philosophy, Computer Science

Project FeederWatch	1989	Open Community Contribution System	Zoology
Theoretical and Computational Biophysics Group (TCBG)	1989	Shared Instrument or Resource	Biomedical Science, Biophysics
Laser Interferometer Gravitational-wave Observatory (LIGO)	1990	Shared Instrument or Resource	Astrophysics, Nuclear Physics

It is interesting that eleven of these are still very active. The three that have ended (ATBC, CARET, PORT) still have websites that offer information or even specimens. Some (LIGO, Ocean.US, and TCBG) started as collaborative projects, and now in their maturity are maintained resources, more like Virtual Communities of Practice. There is now an opportunity to track the lessons learned from these sustained enterprises to feed back into the recommendations here.

A few other interesting observations we have gleaned from the process of filling in and adding new collaboratories are as follows. Again, given that there are still several hundred more collaboratories to update, these findings are preliminary, yet lay the groundwork for continued research and are interesting to note.

- Several large and successful collaboratories have recommended structures in place such as external scientific advisory boards, partitioning of the research into work packages or working groups, and a governing steering group.

- A new category of collaboratory has emerged in recent years: "collaboratory of collaboratories." These "meta-collaboratories" add a new level of collaboration by serving as coordination centers for networks of collaboratories working in the same domain. We have noted 33 of these so far.

- In addition to the 43 collaboratories designated as Community Infrastructure Projects, there are a 41 single organization (non-collaborative) projects that are building tools expressly to support collaborative research. We are compiling a list of these offline.

There are opportunities to continue populating the SOC database, and in the future update the kinds of information highlighted in this chapter.

CHAPTER 11

The Collaboration Success Wizard

The data collection we did in developing the theory of remote scientific collaboration (TORSC) was done primarily through interviews, either in person or over the telephone. In order to expedite future data collection as well as make our findings accessible to current and future collaboratories or collaborations, we have created an on-line survey that we call the Collaboration Success Wizard (CSW). In addition, we provide the person who fills out the survey a personal report on where the collaboration appears to be in good shape, where it might be vulnerable, and what to do about the vulnerabilities. This advice is culled both from our experience with a myriad of best practices and from recommendations for success in the literature to. Additionally, we offer the service of providing the principal investigators or managers of the collaboration a composite report, again of strengths, vulnerabilities and remedies. We provide information at the end of this chapter on how to participate.

11.1 DETAILS OF THE WIZARD

Although our theory originally focused on collaborations in scientific research, almost all of the questions are relevant to most other kinds of distributed collaborations. There is an opportunity to extend the CSW to other domains of collaboration, such as corporations and non-profit organizations. There is additionally the opportunity to develop a version of the CSW that will allow us to explore different forms of organization for collaborations, such as consortia that provide some form of infrastructure or resource for a collection of projects.

The CSW operates like many online surveys, with a few key differences. First, while the survey is administered to individuals, our unit of analysis is usually the project. We work with a coordinating contact on each target project to obtain as extensive a list of individuals involved in the project as possible. We intentionally work to include both central and peripheral members, and to survey people in all roles in the project.

We currently have three versions of the CSW designed for projects in three different stages. For new projects that are just starting, we provide a version worded in future tense that informs about participants' expectations for the project. With ongoing projects, we provide a present-tense version that gives a project an opportunity to pause and reflect on how the collaboration is going and adjust to changing circumstances. For recently completed collaborations, we have a past-tense version that has been used both to understand what happened in the project and provide "lessons-learned" for future projects.

We have a website that describes the Wizard and contains a form through which a person can apply to access the Wizard (hana.ics.uci.edu/wizard). We work with the project leaders to identify projects for which the CSW makes sense.[29] Once the case is approved, we get from the project leaders a list of the names and email addresses of all the participants. We work with them to determine the appropriate timing for the survey and a plan for inviting participants. We frequently ask a senior project leader to send an announcement to all participants in order to encourage maximum participation, but the actual survey invitation will come directly from us to ensure participants that their individual responses will remain confidential. We also send reminders at set intervals to participants who have not yet completed the survey. Depending on the time of year, the size of the project, and other factors, the survey period for each project is usually two to four weeks.

After participants log in to the survey, they are presented with a study information sheet detailing their participation and the data we collect. Participants click through to begin the survey. Each question is presented on a single page. Figure 11.1 shows an example of a page from the version that focuses on a future project. A navigation tree along the left side of the screen shows one's progress through the survey and enables respondents to re-visit prior questions if they wish to reconsider their responses. Participants have the option of signing off and returning later if they cannot complete it in one sitting. Each question is accompanied by a free-text field to explain a response (if their situation does not match the question well), to provide examples, and to make any comments they would like to make, either about the substance of the question or about the comprehensibility of the question itself. Several questions employ conditional branching ("skip logic") such that the next question asked depends on the answer to these questions. We have found that a typical respondent takes about 20-25 min to complete the Wizard depending on how many of the free form text fields they populate with their explanations or color commentary.

---

[29] In our experience so far, only appropriate projects have applied to use the Wizard.

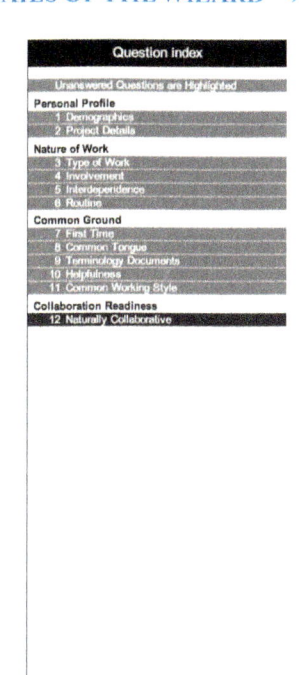

Figure 11.1: A sample question from the Collaboration Success Wizard.

During an interactive interview, it is usually easy to tell if the interviewee understands a question, and if there are problems, the question can be rephrased. However, with an on-line survey, no such interaction is possible. Therefore, we devoted considerable time to iteratively framing and testing our questions, usually by sitting with pilot subjects as they thought out loud while working to complete the Wizard. After our initial, full-scale pilot study, in which we tested the complete CSW, we also interviewed respondents to identify problems and confusions they experienced. The key issue identified was ambiguities of the language used in the survey, typically wanting to know what we meant by certain terms and phrases familiar to us, such as "interdependent," and idioms, such as "common tongue." We began adding clarifying text to questions but, soon, some questions started getting lost in the extra verbiage. This led to the introduction of a separate "note" area for each question, to give examples of what is intended, so that the question itself is distinct.

One important characteristic of the CSW is that when respondents finish the Wizard, they can view a report that provides feedback about the answers they provided. Figure 11.2 shows an example of such feedback. Based on our research and supporting literature, we inform participants about the areas in which they appear to be strong and which areas might require some attention. For the latter, we offer some suggestions of things they could do to mitigate the risk. Considerable effort went into developing the feedback so it would be a smooth as well as informative narrative.

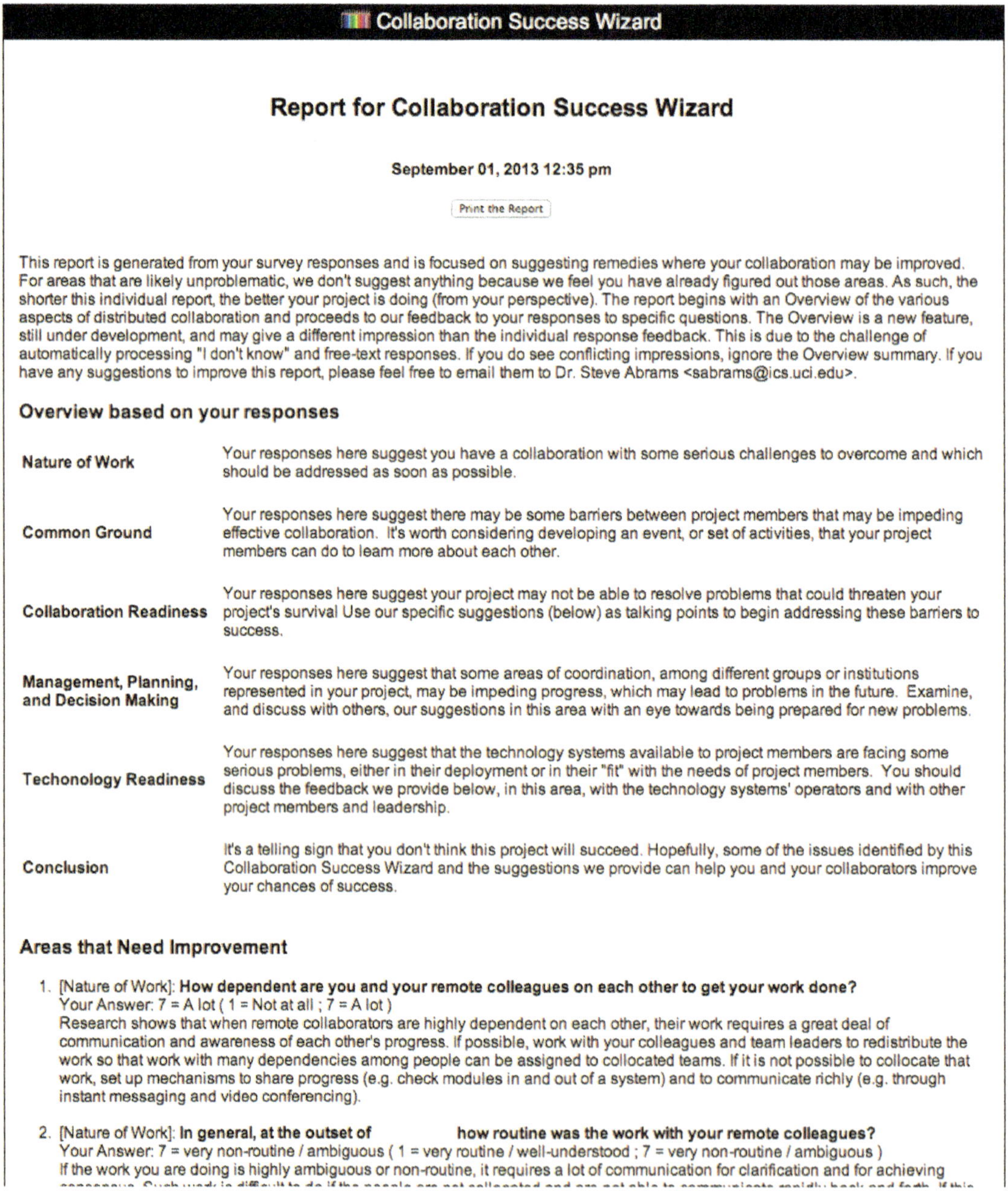

Collaboration Success Wizard

Report for Collaboration Success Wizard

September 01, 2013 12:35 pm

Print the Report

This report is generated from your survey responses and is focused on suggesting remedies where your collaboration may be improved. For areas that are likely unproblematic, we don't suggest anything because we feel you have already figured out those areas. As such, the shorter this individual report, the better your project is doing (from your perspective). The report begins with an Overview of the various aspects of distributed collaboration and proceeds to our feedback to your responses to specific questions. The Overview is a new feature, still under development, and may give a different impression than the individual response feedback. This is due to the challenge of automatically processing "I don't know" and free-text responses. If you do see conflicting impressions, ignore the Overview summary. If you have any suggestions to improve this report, please feel free to email them to Dr. Steve Abrams <sabrams@ics.uci.edu>.

Overview based on your responses

Nature of Work	Your responses here suggest you have a collaboration with some serious challenges to overcome and which should be addressed as soon as possible.
Common Ground	Your responses here suggest there may be some barriers between project members that may be impeding effective collaboration. It's worth considering developing an event, or set of activities, that your project members can do to learn more about each other.
Collaboration Readiness	Your responses here suggest your project may not be able to resolve problems that could threaten your project's survival Use our specific suggestions (below) as talking points to begin addressing these barriers to success.
Management, Planning, and Decision Making	Your responses here suggest that some areas of coordination, among different groups or institutions represented in your project, may be impeding progress, which may lead to problems in the future. Examine, and discuss with others, our suggestions in this area with an eye towards being prepared for new problems.
Techonology Readiness	Your responses here suggest that the technology systems available to project members are facing some serious problems, either in their deployment or in their "fit" with the needs of project members. You should discuss the feedback we provide below, in this area, with the technology systems' operators and with other project members and leadership.
Conclusion	It's a telling sign that you don't think this project will succeed. Hopefully, some of the issues identified by this Collaboration Success Wizard and the suggestions we provide can help you and your collaborators improve your chances of success.

Areas that Need Improvement

1. [Nature of Work]: **How dependent are you and your remote colleagues on each other to get your work done?**
   Your Answer: 7 = A lot (1 = Not at all ; 7 = A lot)
   Research shows that when remote collaborators are highly dependent on each other, their work requires a great deal of communication and awareness of each other's progress. If possible, work with your colleagues and team leaders to redistribute the work so that work with many dependencies among people can be assigned to collocated teams. If it is not possible to collocate that work, set up mechanisms to share progress (e.g. check modules in and out of a system) and to communicate richly (e.g. through instant messaging and video conferencing).
2. [Nature of Work]: **In general, at the outset of how routine was the work with your remote colleagues?**
   Your Answer: 7 = very non-routine / ambiguous (1 = very routine / well-understood ; 7 = very non-routine / ambiguous)
   If the work you are doing is highly ambiguous or non-routine, it requires a lot of communication for clarification and for achieving

Figure 11.2: Sample individual feedback for the Collaboration Success Wizard.

We also offer to provide a project-level summary of what the respondents said through the Wizard about the project. In many ways, this report is similar to the individual report but, with the aggregated data, we can also report statistical trends and distributions reflected in the responses.

Not everyone has the same experience in a collaboration. For example, core members might know more and consequently trust others more than peripheral members. These are important differences to be aware of in managing the collaboration effectively. Just as importantly, we can also analyze the free-text comments provided by respondents, which tend to add a great deal of richness to the numeric data

We have administered the Wizard to 12 projects to date, and they have found this kind of project-level feedback to be very useful. Here are some representative comments, both from projects that were already underway (emphasis added):

- *"...we also assessed our inter-company collaboration effectiveness as a Consortium using the Collaboration Success Wizard, developed by the University of California, Irvine. The results from the assessment **provided direction for changes** in year 3."*

- *"The Collaboration Success Wizard is **an effective management tool** for the NEES Energy Frontier Research Center. It drew out patterns in the way our members work that we were not conscious of, confirmed some impressions we already had, and allowed us to hear frankly from our members. The **Wizard was particularly useful** as an independent evaluation tool that was not tied to a funding agency or other scientific review panel."*

11.2 DETAILS OF THE REPORTS

One of the principal motivations for developing the CSW was to take what we have learned and to help people collaborate across distance more effectively. Both reports acknowledge strengths and suggest strategies for mitigating vulnerabilities, where possible, but the goal of the reports is not to provide a summary judgment. The goal is to facilitate reflection by respondents and to motivate discussion among collaborators. Such consequent discussions are the means by which our theory can be translated into practice and, we believe, by which collaborations can be improved.

Generating a report that can motivate discussion is challenging. There is a limit to the number of questions the CSW can ask and this limits the specificity and detail of the report's feedback. To be most effective, collaborators need to integrate the feedback with their own, more detailed knowledge to understand what it means in their context and to determine what actions they may undertake to improve their collaboration.

Another challenge is that our theory makes no claims about the *relative* impacts of its factors on success. The CSW does include a scoring mechanism, however, to generate an assessment for each category of factors (see Table 4.1) in the report. As we collect more data, however, we expect some factors will emerge as more predictive of success and feedback can be modified to emphasize these aspects.

At present the use of the Wizard is free because it is federally grant-supported.

11.3 INITIAL EXPERIENCE WITH THE WIZARD

We have administered the CSW to 12 projects so far. First we give two illustrations of the impact of the CSW on projects, then give some general findings of the five projects we have analyzed in some detail.

Two examples illustrate the impact of CSW: one project we worked with was a group associated with the University of California Medical Center, who were in the process of applying to the National Institutes of Health (NIH) for a Clinical and Translational Science Award (CTSA). As this was a proposal to create such a center, we administered the future version of the Wizard. We were given a list of 69 names to invite to participate, 43 of whom completed the Wizard survey for a 62% response rate. We prepared an oral and written summary of the results for the CTSA steering committee, who then put information from the summary into their proposal to NIH. They were subsequently successful in obtaining an award and, in the fall of 2010, launched the Institute for Clinical and Translational Science (ICTS) at UC Irvine. There is now an opportunity to follow up with this organization to see how well the Wizard did in characterizing the project when it was in its proposed phase and what their responses were to our feedback.

Another collaborative project, the Collaboration Consortium, is a multinational industry workgroup, sponsored by Cisco Systems that was chartered in 2008 to develop a Collaboration Framework™ to be deployed and rigorously evaluated in their own organizations. As an ongoing collaboration among 14 organizational representatives, they used the present version of the Wizard to assess their workgroup's own collaboration. Nine of the 14 representatives completed the CSW. In their 2011 annual report,[30] their Wizard "experience" and our findings are summarized, as well as the changes they plan to make, which resulted from our analysis and their discussions of our findings. The Wizard afforded them the opportunity to reflect on their collaboration and redirect some aspects, which TORSC identifies as a factor contributing to successful collaboration (Olson et al., 2008).

Across the 12 projects that have used the CSW, we have collected more than 200 completed responses. One project had too few responses to be representative (prompting a concerted effort to better understand what we can do to motivate participation once the leaders have deemed it important to assess the project). Here we summarize some of the themes across the 5 projects we have analyzed in-depth to date.

Common Strengths. Four of the five projects had very high levels of respect for the project manager or core leaders. People were generally thought to be collegial, helpful and collaborative (with one notable exception below). Two mentioned explicitly that they had a feeling of group self-efficacy, the ability to overcome obstacles as they arose. One of the projects with the highest self rating of success was high on all three kinds of trust (that others are responsive, do good work and

[30] communities.cisco.com/docs/DOC-26054

will look out for their interests). Their goals were aligned, unlike many of the weaknesses reported in the other projects. Two were high on recognized good communication plans, with planned face-to-face meetings, regular conference calls, and good use of email.

Common Challenges. Unfortunately, there were many more aspects in common in the list of project challenges. Two projects had clear issues with goals not being aligned (in one case, respondents reported that some people gave and others were "free riders," happy to receive but not willing to give). All had concerns about some aspect of trust, primarily about not being able to rely on those who were remote, and in one case severe loss of personal trust. That is, in one project, it was clear that in general respondents were mixed about how collaborative and trustworthy their colleagues were, and one person, likely the perpetrator of the mistrust, rated everything low. In general, some of the mistrust arose from financial distributions that were seen as unfair or uncertainty about how the resources would be distributed or credit assigned. Two projects had outstanding legal issues as well, primarily involving intellectual property

Four of the five projects reported significant issues with vocabulary, including a case of a team with participants whose native language is not English coupled with others using jargon. In one case, a number of participants reported that the vocabulary problem was serious because it was generally unrecognized. The fact that the work was often ambiguous and non-routine (as is true in many new endeavors in research, both in academics and industry) created challenges to communication. Only one of the projects recognized that the easiest part of the collaboration was each working on its own independent "core," but that the difficulties sprung from their integration into a whole.

The surprise is the number of difficulties people report using technologies. Straight technical difficulties arose with the current difficulty in setting up most video conferences, and platform incompatibility for file sharing and calendaring. But more of the difficulties were social habits surrounding the technology. Most of the discomfort arose in email from not getting responses when expected, and having difficulty arranging times for synchronous communication.

Many of the other challenges had to do with uncertainty about direction, support, engagement of remote participants, and unclear determinants of decisions. This underscores the challenge of distance to management; while many things are transparent when people are collocated, everything takes more explicit effort for clarity when people are distributed.

11.4 THE WIZARD AS TRANSLATIONAL RESEARCH

The CSW is an HCI example of translational research; we wish to bring the results of our research into practice. Such translational research is a major theme in medicine, as evidenced by NIH's creation of over 60 Clinical Translational centers, like the one we studied at the University of California. Translational research frames an important challenge for HCI researchers. In some cases, there is a clear route. For example, a new interaction technique may appear in a commercial product or researchers develop a technical toolkit that makes other developers' work easier. In other cases like participatory design and action research, research and practice are tightly linked through direct interaction. Translating theory into generalizable practical interventions, especially in richly socio-technical or organizational aspects of HCI, remains difficult. But our experience with the CSW suggests that the effort can be fruitful.

The additional benefit of providing the CSW to a wide audience, all of whom can be helped by the analysis, is that we as researchers then can aggregate large amounts of data. These data can in turn generate more findings, such as which of the myriad of factors we ask about are the most critical, and which trade off with each other. Doing translational research, if tightly coupled with research itself, can be a win-win. For more discussion of this, go to (Bietz et al., 2012).

11.5 CONTACT RE THE WIZARD

If the Wizard seems like something that would be useful for assessing your project, go to www.hana.ics.uci/wizard/ for more information and a form for applying to use the Wizard. Also feel free to contact the authors of this volume (jsolson@uci.edu, gary.olson@uci.edu).

CHAPTER 12

Summary and Recommendations

In the previous chapters we described much of what is known in the literature and from our own observations about the factors that make distance work difficult.

Simply put, if the distributed team:

- assigns independent work modules to locations so that infrequent communication done through lean media is sufficient;

- is made up of people who have worked together and have common ground and work styles;

- who like working together;

- who adopt an explicit management style that makes decision making clear, promotes an open, inclusive atmosphere, and has details worked out; and

- use technologies to support communication, coordination, the sharing of data/knowledge, and is supported by an infrastructure powerful enough (both networking and computation),

then it is likely that this team will succeed. This situation is rare.

A distributed team is more likely to:

- be working on new things for which require constant communication to work out the details;

- be made up of people from different backgrounds, both to get diverse ideas generated and to fit the product or service to the people it is designed for;

- be put together by management or dictate from a funder;

- be casual about the specifics of management and oblivious of the important kinds of information they are no longer privy to because other members are "out of sight" and therefore "out of mind;" and

- choose technologies they are used to, like email, missing the opportunities afforded by others (like shared documents and calendars, standardized repositories for shared data), and not supporting rich communication technologies when they are most needed.

These teams will encounter stresses that interfere with success, and, sadly, either meet the challenges or fail.

In what follows, we first list the kinds of general stressors distributed teams encounter and then address the practical recommendations this research suggests. We provide recommendations at three levels: To the individual who is a member of a distributed team, to the manager of a distributed team, and to the institutions that support distributed teams. These recommendations are our attempt to provide research findings that are useful.

12.1 WHAT ABOUT DISTANCE MATTERS?

Many of the implications of the preceding chapters about the nature of work, common ground, collaboration, management, and technology support are simply "good management." What is particularly difficult about work where people are in different locations make these recommendations special?

12.1.1 BLIND AND INVISIBLE

When working with people at distant locations, team members and managers are invisible to those at other locations. And, symmetrically, individuals are blind to the actions and situations of the distant others. This means that all necessary information must be communicated explicitly. As described in the chapter on Technology Readiness, this means both deliberate communication through heavy use of email or blogs or wikis (asynchronous) or chat, and audio or video conferencing (synchronous). And, coordination must be explicitly arranged; it is almost impossible to have a spontaneous meeting, one that would, for example, discuss next actions in response to some unexpected event.

Second, people working with remote colleagues often are unaware of the detailed context of current work. Without explicit communication, remote others do not know what individuals are working on, what their roadblocks and challenges are, and how they can help or be helped. There is no "looking over one's shoulder." All of the work awareness must be tracked explicitly, requiring both an explicit act of recording (like putting activity in a repository that tracks who is doing what, like GitHub, or using Google Docs with revision history) and looking at the record to see what has changed recently and who changed it.

There are additional issues of awareness not about the details of work but about the higher-level conditions of work. We mentioned two of these instances previously: the meeting the manager scheduled was during a blizzard at one location, making it hard for those people to either participate or request a reschedule; and another meeting that continued long into the time which the remote location (France) normally did not work (Friday afternoon). Conversations that include people at the "home" location (headquarters or often the location with the largest number of people) blithely include references to weather, politics, and sports familiar to their local participants but not to those in remote locations, creating an implicit imbalance of power and status because those at the home location are seen as part of an in-group, those remote are the out-group.

12.1.2 TIME ZONE DIFFERENCES

Often, working with people in different locations involves time zone differences. It can range from being an hour off to having no overlap in people's working days. The former merely requires calculations to accommodate the inconvenience of interfering with one's lunch or day's end; the other requires difficult scheduling and someone being inconvenienced.

12.1.3 CROSSING INSTITUTIONAL OR CULTURAL BOUNDARIES

In science, many large-science projects involve a number of organizations, each of which is likely to have its own institutional procedures and regulations, down to the simple differences in the Institutional Review Board policies to severe, show-stopping differences in how intellectual property is viewed. In the corporate world, many long-distance teams reside in the same institutions, where the rules are the same, but the local interpretations might differ. In the non-profit world, it is common to have a mix of distant people within the organization but then partnered with local civic organizations, which also have different rules and expectations.

Of course, very long-distance collaborations then also often involve crossing national cultural boundaries, which have even more striking differences and violations of expectations.

12.1.4 UNEVEN DISTRIBUTION AND THE CONSEQUENT IMBALANCE OF POWER OR STATUS

Although not universal, when there are distributions of people such that there are some locations with only 1-2 people, the small locations often enjoy less status or power. The location with the larger number of people is often "headquarters," where the culture dominates and because the remote people are few, the fact that they are "out of sight, out of mind" reduces their perceived status and therefore power (O'Leary and Cummings, 2007).

12.2 RECOMMENDATIONS

In what follows, we outline the recommendations for people who are on teams that have remote members. In the first section, the recommendations are targeted to individuals who are on distributed teams. The second section is for managers of distributed teams. The third is for people in a role in an institution where they affect policy about infrastructure, rules of engagement, etc. All of these are built on the literature reviewed in the previous 11 chapters, summarized most succinctly in Table 4.1 in Chapter 4. In each section, we group the recommendations around the five major clusters of factors, as appropriate.

12.3 RECOMMENDATIONS CONCERNING THE INDIVIDUALS WHO ARE MEMBERS OF A DISTRIBUTED TEAM

12.3.1 COLLABORATION READINESS

Individuals who work best in distributed teams are those who on personality inventories score high as extroverts and have a social intelligence that suggests they will monitor and respond appropriately to actions and attitudes of others on their team. Individuals who do not have these traits or behaviors are best to recognize the difference and seek assistance in dealing with others. An additional important behavior is being trustworthy. Trust is an important binder of a team, especially those who have infrequent contact. Therefore, behaving in a way that engenders trust is vital.

Another key factor to the success of distributed teams is the motivation of individuals to work with each other. Sometimes the motivation is intrinsic; some of the successful collaborations, like Zebrafish Information Network (ZFIN), involved people who were schooled together, had a common cause in honoring their advisor, and liked working together. Motivation could come from believing in the goal of the project; it could come from incentive structures set by the higher organization, which involve credit on performance review. This latter factor involves decisions on policies determined at the organizational level, discussed later.

12.3.2 TECHNICAL READINESS

Working on a distributed team involves communication and coordination through technology, like email, audio/video conferencing, and more sophisticated technologies for scheduling and coordinating output, such as documents or data. There are two components to this: a disposition to learn new technologies and the training appropriate to make the learning easy. At the individual level is the openness to explore new ways of working that make explicit actions that one normally doesn't have to think about. And then the individual has to commit the time to learn the new technologies, both to get started and then to ask and share best practices as the technology is adapted to the work. At the organizational level, of course, this also involves choosing the technology suite (see Chapter 9) that suits the work, is easy to use, and is compatible across applications.

12.4 RECOMMENDATIONS FOR THE MANAGER OF A DISTRIBUTED TEAM

12.4.1 SELECTING PEOPLE FOR THE TEAM

In the corporate and non-profit worlds, teams are put together primarily by a manager trying to get the right expertise on the team, secondarily because someone is a good team player, being extroverted and with a good social intelligence, described above. Trustworthiness might or might not play into the selection decision, if indeed this is known. In academia, individuals either self-select into a team or are asked by other researchers to collaborate. As in the corporate and nonprofit worlds, most likely this is because of a particular area of expertise, not because they are particularly good at being a team player. A recommendation following realization of how important these individual factors are to team success might be to both assess these factors and put together the team according to expertise and personality traits.

12.4.2 COMMON GROUND

Because the individuals are from different locations, their experience bases are likely to be more different than those individuals who are collocated. This is both good and bad. If they have worked together successfully in the past, they are likely to have worked out vocabulary and working style issues. If not, then it is recommended that they engage in activities where establishing common vocabularies and work style are explicit goals. Working through a Covenant relevant to these topics is recommended. This is especially important if the team members come from different institutions and/or cultural backgrounds. Explicit assessment of habits and expectations and follow-on discussions of difference and ways to resolve differences is highly recommended. As mentioned earlier, assessment through GlobeSmart™ can provide both assessment and guidance.

If the team involves members from different disciplines, or sometimes even different sub disciplines, it is important to work on shared vocabulary and concepts. We reported earlier on examples where a dictionary of terms was developed early so such differences could be overcome. Using the same words to mean different things, or not even knowing some terms, can easily derail a project.

12.4.3 COLLABORATION READINESS

Individuals on a distributed team may have intrinsic motivation to work together, either through personal ties or realization that they need each other's expertise in order to succeed. Both of these behaviors generate respect; when people feel they are respected, they are more likely to be motivated to contribute. If these conditions don't hold, then it is recommended that the manager design explicit motivators for the team, including group rewards and individual incentives that reflect how well the person contributed to this team. Some people adopt a "360° review," in which individuals are evaluated not only by those who manage them, but also by their colleagues and those who work for them.

Two other activities are recommended for managers to bolster the chance of their teams' success. First, since trust is slow to develop in a distributed team (there are fewer occasions for people to get information about how trustworthy someone is, and the ancillary activities of getting familiar with a person's personal life are fewer), managers should generate occasions for exercises or activities for developing trust. In the corporate world, this sometimes involves ropes courses where team members help each other through physical tasks. It could also involve less elaborate sessions where people are encouraged to talk about their non-work lives, where they share things about themselves that indicate vulnerability, an essential ingredient to trust. This is one of the primary reasons to have a face-to-face meeting of all participants at the outset of a project.

The second, related team attitude that helps ensure success is group self efficacy, an attitude of "we can do it." This attitude provides incentives for people to do extra work/contributions when obstacles arise. Again, team-building exercises can help engender this attitude.

12.4.4 THE NATURE OF THE WORK

Whether the work is routine or not (where people know what to do and what others are doing to coordinate their work) is one determinant of whether the distance collaboration is going to be successful or not. Second, if the work can be divided into modules so that most of the coordination and discussion happens among people who are collocated, success is more likely. In a hierarchical organization, the role of designing the work to fit the locations falls to the manager. In less structured organizations, like academia and some non-profit organizations, this recommendation falls to the team members themselves, their project manager, if there is one, and/or their oversight board. Because of the stresses of distance to awareness, communication and coordination, the design of the work is critical. The more that essential communication is required across locations, the less likely the collaboration will be a success.

12.4.5 MANAGEMENT

Most of the category on Management and Decision Making are prescriptions for the manager of a distributed team to help the team achieve success (see Chapter 8 for details). To highlight some of the more important ones, we recommend a number of explicit activities. Helping the individuals come to agreement about their goals, agreeing to the overarching group goal while acknowledging the secondary individual goals is paramount. Aligning the reward structure to fit the group goal additionally ensures that people will understand the priorities of the project and how their behavior affects their personal rewards. Developing an explicit management plan (roles and responsibilities) and outlining how the team is expected to communicate and coordinate helps the team members act appropriately. Having the decision-making process open, free of favoritism, and inclusive both engenders trust and encourages motivation to contribute to the goals.

One of the important management activities for distributed teams occurs in meetings. First, wherever the manager resides is often considered the seat of power, especially if a majority of the team members are collocated there. Meetings present a challenge because of the unreliability of video/audio conferencing, and the lack of cues about who would like to speak next or people's reactions to what is being said. The manager must explicitly solicit commentary and contributions from everyone, even polling locations for commentary. Not only does this ensure that needed information and opinions are heard, but also those at the lesser locations feel respected for being asked. Also, with meetings among people who reside in disparate time zones, it is important that the manager make fair the inconvenience of working outside of regular work hours to participate in the real-time meeting.

It is essential that the manager be proactive in finding out what people are up to. In a collocated setting, this is done by management through walking the hallways. In a distributed team, this requires regular contact with all team members. Frequent e-mail checks, IM chats, voice or video contacts, are all critical. This also helps team members know that they are valued members of the collaboration.

12.4.6 TECHNOLOGY READINESS

Part of putting together a skilled team is selecting those who are comfortable with existing technology and willing to learn and adapt work practices to new technologies. In addition, providing adequate training and ongoing support will make the use of the required technologies easier. There are also resource issues in providing the technology necessary for successful distance work—the shared repositories, the video-conferencing, the shared calendars, etc. that have the right price and the required security. As noted in Chapter 9, non-profits may have the hardest time with the resource issues, and corporations with the security issues.

12.5 RECOMMENDATIONS FOR AN ORGANIZATION THAT WISHES TO SUPPORT A DISTRIBUTED TEAM

A number of the issues faced by the manager of a distributed team fall at the organizational level. For example, the incentive structures a manager can use are often dictated by the organization. The culture of collaboration/competition is often a result of the entire organization or even profession, not in control of the manager. And, often the project design, designating how many people are at each site is dictated by the organization, not the individual manager. And, budget for technical capabilities and support is often dictated by the project budget, which is ultimately determined by the funding agency or the organization. It is incumbent on the project manager to argue for the importance of these factors, but often it is the keeper of the funds who makes the final allocation.

When multiple organizations are involved, as if often the case in long distance collaborations, then there are additional issues to work out. It is at this level that we see explicit activities around aligning the goals of the project, fitting the institutional-specific goals into a secondary role. There may be legal issues that have to be negotiated, and financial issues as well (even down to the distribution of allocated funds being done differently in different countries). In academia, in large projects, there is the matter of who gets credit for the results, not just the publications, but at the organizational level, who gets credit for the funding award and who owns the intellectual property.

12.6 IN THE FUTURE, WILL DISTANCE STILL MATTER?

As we noted in Chapter 1, the Institute for Corporate Productivity reports that 67% of companies with 10,000 or more employees anticipate increasing their use of virtual teams. And, 35% of the respondents rate management of these teams as a top challenge. Because so many people are experiencing distance work, there has been a great deal of progress in the last few years in each of the key areas.

More managers are realizing the success in dividing the work and assigning it to locations so that there is less of a need for heavy communication and coordination across sites. And, we see an increasing discussion of the value of having face-to-face "kick-off meetings" at the start of a project so that discussions about goals and operating procedures are easy and that in the off-times, socializing happens that builds trust. If teams are multi-disciplinary, some attention is paid to developing a common vocabulary, and creating an open atmosphere so that people are encouraged to ask if they don't understand something.

We have also seen more attention paid to crafting the incentives so that individuals are rewarded not only for their own performance but also for the performance of the team. This encourages people to assess how the team is progressing and look for opportunities to help those who are behind. Some organizations are investing in team exercises that engender trust, like ropes courses and exercises where the team builds something physical together collaboratively. Again, these sessions not only build trust, but also help people build a sense of helping others succeed.

On the academic front, funders are asking for management plans, and some large projects are hiring project managers, people skilled more in the management of the project than in the content. There is a growing market for books about how to manage virtual teams (e.g., *Virtual Teams that Work* (Gibson and Cohen, 2003) and *Virtual Team Success* (DeRosa and Lepsinger, 2010)).

And, where teams cross cultural boundaries, some teams are assessing their perceptions and assumptions through "GlobeSmart" assessment tools, and engaging in team activities that reveal differences and work through ways to find a middle ground. For teams that include people in wildly different time zones, some adjust their work schedule so that synchronous communication is possible.

On the technology front, there are more ways for people to be online even though they are mobile, creating both awareness and common ground. Knowledge management is made easier with Sites, wikis, Google docs, and Dropbox. People have developed and adopted a number of applications for awareness support through social computing sites like Facebook, chat, and free video like Skype and Hangouts. Google reports success on some distributed teams, for example, keeping Hangouts open continuously, mimicking the research prototype of Portholes from the 1990s (Dourish and Bly, 1992).

In spite of all these advances and increasing awareness of the difficulties in distance work, distance still matters. Four related areas stand out: (a) being invisible and blind; (2) having to develop trust; (3) crossing cultures; and (4) accommodating non-overlapping work hours.

Without open video connections, where the distributed team members are no longer blind and invisible, it is too easy to forget about them. Out of sight, out of mind. Without co-presence or open video, all communication must be deliberate. In our own experience, even being experts in this area, we come to a meeting that includes remote participants with documents to pass around, forgetting to send them to the remote people. We then take time to do that in the meeting and incur the inevitable delay. When people are on speakerphones, they have trouble indicating they'd like to speak, and we forget to ask them to contribute.

Since trust is developed through interactions both about work (being trustworthy in getting the work done on time and with high quality) and about personal life (finding that others have the same out-of-work stressors, challenges, and joys that you do), trust is hard to develop long-distance. Like being invisible and blind, trust takes explicit effort. It doesn't come free like it does when one is collocated.

Working with people from other cultures continues to be a challenge for a number of reasons. Some individuals who are low on the personality factor of social intelligence may just not see the differences and continue to act in ways that create difficulties. Even those who do have high social intelligence, understanding others who have very different perceptions and habits require effort and knowledge about how to come to a common understanding or appropriate action.

Teams made up of people in very different time zones have smaller windows of time to communicate in important ways synchronously. Someone needs to accommodate in order to create some time for clarification and sharing. The pressures are also often accompanied by power differences, with the lower status location having to accommodate, incurring the inconvenience.

So, the simple answer is that in spite of the important advances in technology and management and participation practices, distance will still matter, but not as much. The effort to accommodate to all these stressors is worth it, and in time it will get easier. And we should not lose sight of the important dictum of Hollan and Stornetta (1992), who argued that technology may make possible things that are impossible in face to-face settings yet have considerable value to collaborative activities.

References

(AIP), A. I. o. P. (1992). *Study of Multi-Institutional Collaborations: Phase I: High-Energy Physics.* New York, American Institute of Physics. 18

Abbott, A. (2002). "Alliance for cellular signaling: Into unknown territory." *Nature* 420: 600-601. DOI: 10.1038/420600a. 8

Acker, S. and S. Levitt (1987). "Designing videoconference facilities for improved eye contact." *Journal of Broadcasting & Electronic Media* 31(2): 181-191. DOI: 10.1080/08838158709386656. 62

Ackerman, M. S., E. C. Hofer, et al. (2008). The National Virtual Observatory. *Scientific Collaboration on the Internet.* G. M. Olson, A. Zimmernan and N. Bos. Cambridge, MA, MIT Press: 135-142. DOI: 10.7551/mitpress/9780262151207.003.0008. 2

Ackerman, M. S. and T. W. Malone (1990). *Answer garden: A tool for growing organizational memory.* COIS '90, New York, ACM Press. 72

Adamic, L. and N. Glance (2005). *The political blogosphere and the 2004 U.S. Election: Divided they blog.* LinkKDD Workshop. 63

Allen, T. (1977). *Managing the Flow of Technology.* Cambridge, MA, MIT Press. 1

Archambault, A. and J. Grudin (2012). A longitudinal study of Facebook, LinkedIn, & Twitter use. *CHI 2012.* New York, ACM: 2741-2750. DOI: 10.1145/2207676.2208671. 29

Argyle, M. and M. Cook (1976). *Gaze and Mutual Gaze.* New York, Cambridge University Press. 62

Atkins, D. E., K. K. Droegemeier, et al. (2003). *Revolutionizing Science and Engineering Through Cyberinfrastructure. Report of the National Science Foundation Blue-Ribbon Advisory Panel on Cyberinfrastructure.* Washington, DC, National Science Foundation. 12, 45

Bainbridge, W. S. (2007). "The scientific research potential of virtual worlds." *Science* 317: 472-476. DOI: 10.1126/science.1146930. 65

Bandura, A. (1977). "Self-efficacy: Toward a unifying theory of behavioral change." *Psychological Review* 84(2): 191-215. DOI: 10.1037/0033-295X.84.2.191. 47

Bennett, L. M., H. Gadlin, et al. (2010). *Collaboration and team science: A field guide.* Bethesda, MD, National Institutes of Health. 3

Berman, F., P. E. Bourne, et al. (2004). "The protein data bank: A case study in management of community data." *Current Proteomics* 1(1): 49-57. DOI: 10.2174/1570164043488252. 55

REFERENCES

Berman, H., K. Henrick, et al. (2003). "Announcing the worldwide Protein Data Bank." *Nature Structural Biology* 10(12): 980. DOI: 10.1038/nsb1203-980. 1

Bernstein, F. C., T. F. Koetzle, et al. (1977). "The Protein Data Bank: A computer-based archival file for macromolecular structures." *Journal of Molecular Biology* 112: 535-542. DOI: 10.1016/S0022-2836(77)80200-3. 1

Beyer, H. and K. Holtzblat (1998). *Contextual Design*. San Francisco, Morgan Kaufmann Publishers. 81

Bietz, M., M. Naidoo, et al. (2008). International AIDS Research Collaboratories: The HIV Pathogenesis Program. *Scientific Collaboration on the Internet*. G. M. Olson, A. Zimmerman and N. Bos. Cambridge, MA, MIT Press: 351-363. 27, 73, 74

Bietz, M. J., S. Abrams, et al. (2012). "Improving the odds through the Collaboration Success Wizard." *Translational Behavioral Medicine* 2(4): 480-486. DOI: 10.1007/s13142-012-0174-z. 104

Bikson, T. K., G. F. Treverton, et al. (2008). Leadership in international organizations: 21st century challenges. *Leadership at a Distance: Research in Technologically-Suppported Work*. S. P. Weisband. New York, Lawrence Erlbaum Associates: 13-30. 51

Birnholtz, J. (2008). "When authorship isn't enough: Lessons from CERN on the implications of formal and informal credit attribution mechanisms in collaborative research." *Journal of Electronic Publishing* 11(1). DOI: 10.3998/3336451.0011.105. 51

Birnholtz, J. and S. Ibara (2012). *Tracking changes in collaborative writing: Edits, visibility and group maintenance*. CSCW 2012, New York, ACM Press. DOI: 10.1145/2145204.2145325. 78

Birnholtz, J. P. (2006). "What does it mean to be an author? The intersection of credit, contribution and collaboration in science." *Journal of the American Society of Information Science and Technology* 57(13): 1758-1770. DOI: 10.1002/asi.20380. 37

Blackburn, R. S., S. A. Furst, et al. (2003). Building a winning virtual team. *Virtual Teams That Work: Creating the Conditions for Virtual Team Effectiveness*. C. B. Gibson and S. B. Cohen. San Francisco, Jossey-Bass: 95-120. 43

Bonney, R., C. B. Cooper, et al. (2009). "Citizen science: A developing tool for expanding science knowledge and scientific literacy." *BioScience* 59(11): 977-984. DOI: 10.1525/bio.2009.59.11.9. 14, 16, 27, 75

Borghese, R. J. and P. F. Borgese (2002). *M&A from Planning to Integration: Executing Acquisitions and Increasing Shareholder Value*. New York, McGraw-Hill. 39

REFERENCES

Borning, A., B. Friedman, et al. (2005). *Informing Public Deliberation: Value Sensitive Design of Indicators for a Large-Scale Urban Simulation*. ECSCW 2005, Springer. DOI: 10.1007/1-4020-4023-7_23. 2

Bos, N. (2008). Motivation to contribute to collaboratories: A public goods approach. *Scientific Collaboration on the Internet*. G. M. Olson, A. Zimmerman and N. Bos. Cambridge, MA, MIT Press: 251-274. 14, 17, 44, 45

Bos, N., J. S. Olson, et al. (2002). Effects of four computer-mediated communication channels on trust development. *CHI 2002*, New York, ACM Press. DOI: 10.1145/503376.503401. 46

Bos, N., N. S. Shami, et al. (2004). *In-group/out-group effects in distributed teams: An experimental simulation*. CSCW '04, ACM. DOI: 10.1145/1031607.1031679. 50

Bos, N., A. Zimmerman, et al. (2008). From shared databases to communities of practice: A taxonomy of collaboratories. *Scientific Collaboration on the Internet*. G. M. Olson, A. Zimmernan and N. Bos. Cambridge, MA, MIT Press: 53-72. 7, 23

Boulos, M. N. K., L. Hetherington, et al. (2007). "Second life: An overview of the potential of 3-D virtual worlds in medical and health education." *Health Information & Libraries Journal* 24(4): 233-245. DOI: 10.1111/j.1471-1842.2007.00733.x. 65

Brabham, D. C. (2013). *Crowdsourcing*. Cambridge, MA, MIT Press. 14

Bradner, E. and G. Mark (2001). *Social presence with video and application sharing*. GROUP 2001, New York, ACM Press. DOI: 10.1145/500286.500310. 61

Bradner, E. and G. Mark (2002). Why distance matters: Effects on cooperation, persuasion and deception. *CSCW '02*, New York, ACM. DOI: 10.1145/587078.587110. 52

Bradner, E. and G. Mark (2008). Designing a tail in two cities: Leaders' perspectives on collocated and distance collaboration. *Leadership at a Distance: Research in Technologically Supported Work*. S. P. Weisband. New York, Lawrence Erlbaum Associates: 51-70. 62

Brown, J. S. and D. Thomas (2006). "You play World of Warcraft? You're hired." *Wired* 14(4). 65

Browning, E. S. (1994). Side by side: Computer chip project brings rivals together, but the cultures clash. *The Wall Street Journal* A1, A6. 41

Buxton, W. (1992). *Telepresence: Integrating shared task and person spaces*. Graphics Interface '92, New York, ACM. 62

Cadiz, J. J., A. Balachandran, et al. (2000). *Distance learning through distributed collaborative video viewing*. CSCW 2000, New York, ACM. DOI: 10.1145/358916.358984. 18

REFERENCES

Cai, D. A. and C.-J. Hung (2005). How relevant is trust anyway? A cross-cultural comparison of strust in organizational and peer relationships. *International and multi-cultural organizational communication*. G. C. G. A. Barnett. Cresskill, NJ, Hapton Press: 73-104. 47

Cameron, A. F. and J. Webster (2005). "Unintended consequences of emerging communication technologies: Instant messaging in the workplace." *Computers in Human Behavior* 21(1): 85-103. DOI: 10.1016/j.chb.2003.12.001. 59

Card, S. (2012). Information visualization. *The Human-Computer Interaction Handbook: Fundamentals, Evolving Technologies, and Emerging Applications*. J. A. Jacko. Boca Raton, FL, CRC Press: 515-548. 13

Carley, K. and K. Wendt (1991). "Electronic mail and scientific communication: A study of the Soar extended research goup." *Knowledge: Creation, Diffusion, Utilization* 12: 406-440. 58

Carroll, J. M., M. B. Rosson, et al. (2005). *Collective Efficacy as a Measure of Community*. CHI 2005, ACM. DOI: 10.1145/1054972.1054974. 47

Chamberlin, T. C. (1890). "The method of multiple working hypotheses." *Science* (old series) 15(92). DOI: 10.1126/science.148.3671.754. 2

Chatman, J. A. and K. A. Jehn (1994). "Assessing the Relationship between Industry Characteristics and Organizational Culture." *Academy of Management Journal* 37(3): 522-553. DOI: 10.2307/256699. 44

Chen, Y. and T. Sonmez (2006). "School choice: An experimental study." *Journal of Economic Theory* 127: 202-231. DOI: 10.1016/j.jet.2004.10.006. 70

Chompalov, I., J. Genuth, et al. (2002). "The organization of scientific collaborations." *Research Policy* 31: 749-767. DOI: 10.1016/S0048-7333(01)00145-7. 37

Clark, H. H. (1996). *Using language*. New York, Cambridge University Press. DOI: 10.1017/CBO9780511620539. 39

Clark, H. H. and S. E. Brennan (1991). Grounding in communication. *Perspectives on socially shared cognition*. L. B. Resnick, R. M. Levine and S. D. Teasley. Washington, D.C., American Psychological Association: 127-149. DOI: 10.1037/10096-006. 39, 75, 79

Cooper, S., F. Khatib, et al. (2010). "Predicting protein structures with a multiplayer online game." *Nature* 466: 756-760. DOI: 10.1038/nature09304. 15

Cramton, C. D. (2001). "The mutual knowledge problem and its consequences in geographically dispersed teams." *Organizational Science* 12: 346-371. DOI: 10.1287/orsc.12.3.346.10098. 40

REFERENCES

Cummings, J. N., J. A. Espinosa, et al. (2009). "Crossing spatial and temporal boundaries in globally distributed projects: A relational model of coordination delay." *Information Systems Research* 30(3): 420-439. DOI: 10.1287/isre.1090.0239. 38

Cummings, J. N. and S. Kiesler (2005). "Collaborative research across disciplinary and institutional boundaries." *Social Studies of Science* 35(5): 703-722. DOI: 10.1177/0306312705055535. 3, 49, 50

Cummings, J. N. and S. Kiesler (2007). "Coordination costs and project outcomes in multi-university collaborations." *Research Policy* 36(10): 1620-1634. DOI: 10.1016/j.respol.2007.09.001. 3, 49

Cummings, J. N. and S. Kiesler (2008). *Who collaborates successfully? Prior experience reduces collaboration barriers in distributed interdisciplinary research.* CSCW '08, ACM. DOI: 10.1145/1460563.1460633. 3, 25, 26, 39

Cummings, J. N., S. Kiesler, et al. (2012). "Group heterogeneity increases the risks of large group size: A longitudinal study of research group productivity." *Psychological Science* 24(6), 880-890. 26

David, P. A. and M. Spence (2010). Institutional infrastructures for global research networks in the public sector. *World Wide Research*. W. Dutton and P. W. Jeffreys. Cambridge, MA, MIT Press: 191-213. 54

Davis, D. D. and J. L. Bryant (2003). "Influence at a distance: Leadership in virtual teams." *Advances in Global Leadership* 3: 303-340. DOI: 10.1016/S1535-1203(02)03015-0. 50

de la Flor, G., M. Jirotka, et al. (2010). "Transforming scholarly practice: Embedding technological interventions to support the collaborative analysis of ancient texts." *Computer Supported Cooperative Work* 19(3-4): 309-334. DOI: 10.1007/s10606-010-9111-1. 8

DeRosa, D. and R. Lepsinger (2010). *Virtual team success: A practical guide for working and leading from a distance.* San Francisco, CA, Jossey-Bass. 112

DiSalvio, P. (2012). "Pardon the disruption…innovation changes how we think about higher education." *New England Journal of Higher Education.* 18

Djorgovski, S. G., P. Hut, et al. (2010). Exploring the use of virtual worlds as a scientific research platform: The Meta-Institute for Computational Astrophysics (MICA). *Facets of Virtual Environments.* F. Lehmann-Grube, J. Sablatnig, O. Akan et al. Berlin, Springer. 33: 29-43. 65

Doan, A., R. Ramakrishnan, et al. (2011). "Crowdsourcing systems on the World-Wide Web." *Communications of the ACM* 54(4): 86-96. DOI: 10.1145/1924421.1924442. 75

REFERENCES

Doherty-Sneedon, G., A. Anderson, et al. (1997). "Face-to-face and video mediated communication: A comparison of dialogue strucdture and task performance." *Journal of Experimental Psychology: Applied* 3(2): 105-123. DOI: 10.1037/1076-898X.3.2.105. 61

Donath, J. and F. B. Viegas (2002). *The Chat Circle series: Explorations in designing abstract graphical communication interfaces*. DIS 2002. New York, ACM: 359-369. 59

Dourish, P. and S. Bly (1992). *Portholes: Supporting awareness in a distributed work group*. CHI 92, New York, ACM Press. DOI: 10.1145/142750.142982. 66, 112

Duarte, D. and N. Snyder (1999). *Mastering virtual teams*. San Francisco, CA, Jossey-Bass. 43

Duncan, S. (1972). "Some signals and rules for taking speaking turns in conversations." *Journal of Personality and Social Psychology* 23(2): 283-292. DOI: 10.1037/h0033031. 61

Dutton, W. and T. Piper (2010). The politics of privacy, confidentiality, and ethics: Opening research methods. *World Wide Research*. W. Dutton and P. W. Jeffreys. Cambridge, MA, MIT Press: 223-240. 54

Dutton, W. H. and E. T. Meyer (2010). Enabling or mediating the social sciences: The opportunities and risks of bottom-up innovation. *World Wide Research: Reshaping the Sciences and Humanities*. W. H. Dutton and P. W. Jeffreys. Cambridge, MA, MIT Press: 165-184. DOI: 10.7551/mitpress/9780262014397.003.0020. 8

Ehrlich, K. and N. S. Shami (2010). *Microblogging inside and outside the workplace*. ICWSM 2010, AAAI Press: 42-29. 64

Erickson, T., C. Halverson, et al. (2002). "Social translucence: Designing social infrastructures that make collective activity visible." *Communications of the ACM* 45(4): 40-44. DOI: 10.1145/505248.505270. 59

Falk-Krzesinski, H. J., K. Borner, et al. (2011). "Advancing the science of team science." *Clinical and Translational Science* 3(5): 263-266. DOI: 10.1111/j.1752-8062.2010.00223.x. 3

Feld, S. (1982). "Social structural determinants of similarity among associates." *American Sociological Review* 47: 797-801. DOI: 10.2307/2095216. 46

Forsyth, D. R. (2010). *Group dynamics*. Belmont, CA, Wadsworth. 43

Foster, I. and C. Kesselman, Eds. (2004). *The grid: Blueprint for a new computing infrastructuxre*. San Francisco, Morgan Kaufmann. 74

Freeman, J. C. (2004). A model interdisciplinary collaboration by the Florida Research Ensemble. *Electronic Collaboration in the Humanities*. J. A. Inman, C. Reed and P. Sands. Mahwah, NJ, Lawrence Erlbaum Associates: 335-362. 8

REFERENCES 121

Fuller, R. M. and A. R. Dennis (2009). "Does fit matter? The impact of task-technology fit and appropriation on team performance in repeated tasks." *Information Systems Research* 20(1): 2-17. DOI: 10.1287/isre.1070.0167. 80

Fussell, S. R., S. Kiesler, et al. (2004). *Effects of instant messaging on the management of multiple project trajectories*. CHI 2004, ACM. DOI: 10.1145/985692.985717. 50

Gassman, O. and M. von Zedtwitz (1998). "Organization of Industrial R&D on a Global Scale." *R&D Management* 28(3): 147-161. DOI: 10.1111/1467-9310.00092. 37

Gibbons, J. F., W. R. Kincheloe, et al. (1977). "Tutored videotape instruction: A new use of electronics media in education." *Science* 195: 1139-1146. DOI: 10.1126/science.195.4283.1139. 18

Gibson, C. B. and S. G. Cohen, Eds. (2003). *Virtual teams that work: Creating conditions for virtual team effectiveness*. San Francisco, CA, Jossey-Bass. 112

Golbeck, J., J. M. Grimes, et al. (2010). "Twitter use by the U.S. Congress." *Journal of the American Society for Information Science and Technology* 61(8): 1612-1621. DOI: 10.1002/asi.21344. 64

Golbeck, J. and D. L. Hansen (2011). *Computing political preference among Twitter followers*. CHI 2011. New York, ACM: 1105-1108. DOI: 10.1145/1978942.1979106. 64

Goodhue, D. L. and R. L. Thompson (1995). "Task-technology fit and individual performance." *MIS Quarterly* 19(2): 213-236. DOI: 10.2307/249689. 80

Grayson, D. M. and A. F. Monk (2003). "Are you looking at me? Eye contact and desktop video conferencing." *Transactions on Computer-Human Interaction* 10(3): 221-243. DOI: 10.1145/937549.937552. 62

Grier, D. A. (2005). *When computers were human*. Princeton, NJ, Princeton University Press. 74

Grinter, R. E. (2000). "Workflow systems: Occasions for success and failure." *Computer Supported Cooperative Work* 9: 189-214. DOI: 10.1023/A.1008719814496. 70

Grudin, J. (1994). "Groupware and social dynamics: Eight challenges for developers." *Communications of the ACM* 37(1): 93-104. DOI: 10.1145/175222.175230. 44, 58, 66

Grudin, J. (2004). *Return on investment and organizational adoption*. CSCW 2004, New York, ACM. DOI: 10.1145/1031607.1031659. 31

Grudin, J. and L. Palen (1995). *Why groupware succeeds: Discretion or mandate?* ECSCW 1995, Stockholm, Sweden, Springer. 66

Grudin, J. and E. S. Poole (2010). *Wikis at work: Success factors and challenges for sustainability of enterprise wikis*. WikiSym '10, New York, ACM. DOI: 10.1145/1832772.1832780. 17, 63

REFERENCES

Gutwin, C. and S. Greenberg (1999). "The effects of workspace awareness support on the usability of real-time groupware." *ACM Transactions on Computer-Human Interaction* 6(3): 243-281. DOI: 10.1145/329693.329696. 67

Hackett, E. J., J. N. Parker, et al. (2008). Ecology transformed: The National Center for Ecological Analysis and Synthesis and the Changing Patterns of Ecological Research. *Scientific Collaboration on the Internet*. G. M. Olson, A. Zimmerman and N. Bos. Cambridge, MA, MIT Press: 277-296. 40, 46

Hackman, J. R. (2002). *Leading teams: Setting the stage for great performances*. Boston, MA, Harvard Business School Press. 3

Hackman, J. R. (2011). *Collaborative intelligence: Using teams to solve hard problems*. San Francisco, Berrett-Koehler Publishers. 3

Haines, J. K., J. S. Olson, et al. (2013). *Here or there? How configurations of transnational teams impacts social capital*. Interact 2013. DOI: 10.1007/978-3-642-40480-1_32. 42

Halfpenny, P., R. Procter, et al. (2009). Developing the UK-based e-Social Science Research Program. *e-Research: Transformation in Scholarly Practice*. N. W. Jankowski. New York, Routledge: 73-90. 44

Hall, K. L., D. Stokols, et al. (2008). "The collaboration readiness of transdisciplinary research teams and centers: Findings from the National Cancer InstituteTREC Year-One evaluation study." *American Journal of Preventive Medicine* 35(2 Supplement): S161-S172. DOI: 10.1016/j.amepre.2008.03.035. xiii, 47

Hand, E. (2010). "Citizen science: People power." *Nature* 466: 685-687. DOI: 10.1038/466685a. 14, 15, 27, 75

Hansen, M. T. (2009). *Collaboration: How leaders avoid the traps, create unity, and reap big results*. Boston, MA, Harvard Business Press. 3

Hayes, G. (2011). "The relationship of action research to human-computer interaction." *ACM Transactions on Computer-Human Interaction* 18(3). DOI: 10.1145/1993060.1993065. 29

Heer, J. and M. Bostock (2010). *Crowdsourcing graphical perception: Using Mechanical Turk to assess viisualization design*. CHI 2010. New York, ACM: 203-212. DOI: 10.1145/1753326.1753357. 75

Hemphill, L. and A. Begel (2008). How will you see my greatness if you can't see me? Poster presented at the ACM Conference on *Computer Supported Cooperative Work*. San Diego, CA. 9

Herbsleb, J., D. Zubrow, et al. (1994). "Software process improvement: State of the payoff." *American Programmer* 7: 2-12. 38

REFERENCES

Herbsleb, J. D., A. Mockus, et al. (2000). Distance, dependencies, and delay in a global collaboration. *Computer Supported Cooperative Work*, New York, ACM. DOI: 10.1145/358916.359003. 46, 65

Herlocker, J. L., J. A. Konstan, et al. (2000). *Explaining collaborative filtering recommendations.* CSCW 2000, New York, ACM Press. DOI: 10.1145/358916.358995. 72

Hicks, D., C. Larson, et al. (2008). "The influence of collaboration on programs outcomes: The Colorado Nurse Family Partnership." *Evaluation Review* 32: 453-477. DOI: 10.1177/0193841X08315131. 32

Hinds, P. and D. Bailey (2003). "Out of sight, out of sync: Understanding conflict in distributed teams." *Organizational Science* 14(6): 615-632. DOI: 10.1287/orsc.14.6.615.24872. 46

Hinds, P. and C. McGrath (2006). *Structures that work: Social structure, work structure, and performance in geographically distributed teams.* CSCW 2006, New York, ACM. DOI: 10.1145/1180875.1180928. 49

Hinds, P. and M. Mortensen (2005). "Understanding conflict in geographically distributed teams: An empirical investigation." *Organizational Science* 16(3): 290-307. DOI: 10.1287/orsc.1050.0122. 46

Hislop, D. (2013). *Knowledge Management in Organizations: A Critical Introduction.* New York, Oxford University Press. 13

Hofer, E. C., S. McKee, et al. (2008). High-energy physics: The large Hadron Collider collaborations. *Scientific Collaboration on the Internet.* G. M. Olson, A. Zimmerman and N. Bos. Cambridge, MA, MIT Press: 143-151. 1, 9, 10, 12, 18

Hollan, J. D. and S. Stornetta (1992). *Beyond being there.* CHI '92, New York, ACM Press. DOI: 10.1145/142750.142769. 75, 113

Hooker, J. N. (2012). Cultural differences in business communication. *Handbook of Intercultural Discourse and Communication.* S. F. K. D B. Paulston, E. S. Rangel, Wiley. DOI: 10.1002/9781118247273.ch19. 52

Horvitz, E. and J. Apacible (2003). *Learning and reasoning about interruption.* ICMI '03, New York, ACM. DOI: 10.1145/958432.958440. 67

Howe, J. (2008). *Crowdsourcing: Why the power of the crowd is driving the future of business.* Crown Business. 14, 75

Hu, M., S. Liu, et al. (2012). *Breaking news on Twitter.* CHI 2012. New York, ACM: 2751-2754. DOI: 10.1145/2207676.2208672. 63

124 REFERENCES

Hughes, M. and J. B. Terrell (2012). *Emotional intelligence in action: Training and coaching activities for leaders, managers, and teams.* San Francisco, Pfeiffer. 47

Inman, J. A., C. Reed, et al., Eds. (2004). *Electronic collaboration in the humanities.* Mahwah, NJ, Lawrence Erlbaum Associates. 2, 87

Isaacs, E., A. Walendowski, et al. (2002). *The character, functions, and styles of instant messaging in the workplace.* CSCW 2002, New York, ACM Press. DOI: 10.1145/587078.587081. 59

Ishii, H. and M. Kobayashi (1992). *ClearBoard: A seamless medium for shared drawing and conversation with eye contact.* CHI '92, New York, ACM. DOI: 10.1145/142750.142977. 62

Jankowski, N. W., Ed. (2009). *e-Research: Transformation in scholarly practice.* New York, Routledge. 1

Jansen, B. J., M. Zhang, et al. (2009). "Twitter power: Tweets as electronic word of mouth." *Journal of the American Society for Information Science and Technology* 60(11): 2169-2188. DOI: 10.1002/asi.21149. 63

Jarvenpaa, S. L. and D. E. Leidner (1999). "Communication and trust in global virtual teams." *Organizational Science* 10: 791-815. DOI: 10.1287/orsc.10.6.791. 46

Jirotka, M., R. Procter, et al. (2005). "Collaboration and trust in healthcare innovation: The eDiaMoND case study." *Computer Supported Cooperative Work* 14(4): 369-389. DOI: 10.1007/s10606-005-9001-0. 46

Johnson, W. L. and A. Valente (2009). "Tactical language and culture training systems: Using AI to teach foreign languages and cultures." *AI Magazine* 30(2): 72-83. 65

Johnstone, A., Berry, U., Nguyen, T., Asper, A. (1995). "There was a long pause: Influencing turn-taking behaviour in human-human and human-computer spoken dialogues." *International Journal of Human-Computer Studies* 42(4): 383-411. DOI: 10.1006/ijhc.1995.1018. 61

Kalman, Y. M. and S. Rafaeli (2011). "Online pauses and silence: Chronemic expectancy violations in written computer-mediated communication." *Communication Research* 38(1): 54-69. DOI: 10.1177/0093650210378229. 51

Kaplan, R. S. and D. P. Norton (1996). *The balanced scorecard.* Cambridge, MA, Harvard Business School Press. 30

Katz, L. (2006). *Negotiating International Business—The Negotiator's Reference Guide to 50 Countries Around the World.* Booksurge, L. L. C. 52

Kendon, A. (1967). "Some functions of gaze direction in social interactions." *Acta Psychologica* 26: 22-63. DOI: 10.1016/0001-6918(67)90005-4. 62

REFERENCES 125

Khatib, F., F. DiMaio, et al. (2011). "Crystal strucdture of a monomeric retroviral protease solved by protein folding game players." *Nature Structural & Molecular Biology* 18(10): 1175-1177. DOI: 10.1038/nsmb.2119. 15

Kiesler, S. and J. N. Cummings (2002). What do we know about proximity and distance in work groups? A legacy of research. *Distributed work*. P. J. Hinds and S. Kiesler. Cambridge, MA, MIT Press: 57-80. 35, 51

Kirkman, B. L., C. B. Gibson, et al. (2012). Across borders and technologies: Advancements in virtual teams research. *The Oxford Handbook of Organizational Psychology*. S. W. J. Kozlowski. New York, Oxford University Press. DOI: 10.1093/oxfordhb/9780199928286.013.0025. 49

Kittur, A., E. H. Chi, et al. (2008). *Crowdsourcing user studies with Mechanical Turk*. CHI 2008, New York, ACM. DOI: 10.1145/1357054.1357127. 27, 75

Kittur, A., B. Smus, et al. (2011). *CrowdForge: Crowdsourcing complex work*. UIST 2011, New York, ACM Press. DOI: 10.1145/2047196.2047202. 16, 27

Koehne, B. and D. Redmiles (2012). *Envisioning distributed usability evaluation thorugh a virtual world platform*. CHASE 2012. Zurich, Switzerland: 73-75. DOI: 10.1109/CHASE.2012.6223027. 65

Koehne, B., P. C. Shih, et al. (2012). Remote and alone: Coping with being the remote member on the team. *Computer Supported Cooperative Work*, New York, ACM Press. DOI: 10.1145/2145204.2145393. 9, 49

Koslowski, S. W. J. and D. R. Ilgen (2006). "Enhancing the effectiveness of work groups and teams." *Psychological Science in the Public Interest* 7(3): 77-124. 33

Kraemer, K. L. and A. Pinsonneault (1990). Technology and groups: Assessments of empirical research. *Intellectual teamwork: Social and technical foundations of cooperative work*. J. Galegher, R. Kraut and C. Egido. Hillsdale, NJ, Lawrence Erlbaum Associates: 373-405. 67

Krauss, R. M., C. M. Garlock, et al. (1977). "The role of audible and visible back-channel responses in interpersonal communication." *Journal of Personality and Social Psychology* 35(7): 523-529. DOI: 10.1037/0022-3514.35.7.523. 76

Kraut, R. E., C. Egido, et al. (1990). Patterns of contact and communication in scientific research collaborations. *Intellectual teamwork: Social and technological foundations of cooperative work*. J. Galegher, R. E. Kraut and C. Egido. Hillsdale, NJ, Lawrence Erlbaum Associates: 149-171. 46

126 REFERENCES

Kurland, N. B. and T. D. Egan (1999). "Telecommuting: Justice and control in the virual organization." *Organizational Science* 10(4): 500-513. DOI: 10.1287/orsc.10.4.500. 52

Larson, C., A. Christian, et al. (2002). *Colorado Healthy Communities Initiative: Ten years later*. Denver, CO, Colorado Trust. 52

LeFasto, M. F. J. and C. Larson (2001). *When teams work Best: 6,000 team members and leaders tell what it takes to succeed*. Thousand Oaks, CA, Sage Publications. 52

Levy, S. (2011). *In the plex*. New York, NY, Simon & Schuster. 65

Lewis, L., M. G. Isbell, et al. (2010). "Collaborative tensions: Practitioners' experiences of interorganizational relationships." *Communication Monographs* 77(4): 460-479. DOI: 10.1080/03637751.2010.523605. 2

Ling, K., G. Beenen, et al. (2005). "Using social psychology to motivate contributions to online communities." *Journal of Computer-Mediated Communication* 10(4). 17

Lintott, C. J., K. Schawinski, et al. (2008). "Galazy Zoo: Morphologies derived from visual inspection of galaxies from the Sloan Digital Sky Survey." *Monthly Notices of the Royal Astronomical Society* 389(3): 1179-1189. DOI: 10.1111/j.1365-2966.2008.13689.x. 15

Luo, A. and J. S. Olson (2008). How collaboratories affect scientists from developing countries. *Scientific Collaboration on the Internet*. G. M. Olson, A. Zimmerman and N. Bos. Cambridge, MA, MIT Press: 365-376. 53

Lurey, J. S. and M. S. Raisinghani (2001). "An empirical study of best practices in virtual teams." *Information and Management* 8: 523-544. DOI: 10.1016/S0378-7206(01)00074-X. 30

Mackay, W. F. (1989). "Diversity in the use of electronic mail: A preliminary inquiry." *ACM Transactions on Office Information Systems* 6: 380-397. DOI: 10.1145/58566.58567. 58

Malone, T. W. (2004). *The future of work: How the new order of business will shape your organization, your management style, and your life*. Boston, MA, Harvard Business School Press. 1, 8

Malone, T. W., R. Laubacher, et al. (2010). "The collective intelligence genome." *Sloan Management Review* 51(3): 21-31. 14, 75

Maltz, A. C., A. J. Shenhar, et al. (2003). "Beyond the balanced scorecard: Refining the search for organizational success measures." *Long Range Planning* 36: 187-204. DOI: 10.1016/S0024-6301(02)00165-6. 30, 31

Mark, G. and B. Semaan (2008). *Resilience in collaboration: Technology as a resource for new patterns of acdtion*. CSCW '08, New York, ACM Press. DOI: 10.1145/1460563.1460585. 62

Marshall, C. and J. C. Tang (2012). *That syncing feeling: Early user experiences with the cloud*. DIS 2012, New York, ACM. DOI: 10.1145/2317956.2318038. 13, 54

REFERENCES

Marshall, C. C. and F. M. Shipman (2011). *Social media ownership: Using Twitter as a window onto current attitudes and beliefs*. CHI 2011. New York, ACM: 1081-1090. DOI: 10.1145/1978942.1979103. 64

Mayer-Schoenberger, V. and K. Cukier (2013). *Big data: A revolution that will transform how we live, work and think*. Boston, Houghton Mifflin Harcourt. 54

Mazmanian, M., W. J. Orlikowski, et al. (2005). Crackberries: The social implications of ubiquitous wireless e-mail devices. *Designing ubiquitous information environments: Socio-technical issues and challenges*. C. Sorensen, Y. Yoo, K. Lyytinen and J. DeGross. New York, Springer: 341-347. 60

Maznevski, M. L. and K. M. Chudoba (2000). "Bridging space over time: Global virtual team dynamics and effectiveness." *Organizational Science* 11(5): 473-492. DOI: 10.1287/orsc.11.5.473.15200. 51

Mazur, A. and E. Boyko (1981). "Large-scale ocean research projects: What makes them succeed or fail?" *Social Studies of Science* 11(4): 425-449. DOI: 10.1177/030631278101100402. 50

McCrae, R. R. and P. T. Costa, Jr (2008). The Five Factor Theory of Personality. *Handbook of personality, Third Edition: Theory and research*. R. W. R. Oliver P. John, Lawrence A. Pervin, Guilford Press. 43

McGrath, J. E. (1984). *Groups: Interaction and Performance*. Inglewood Cliffs, NJ, Prentice-Hall. 31, 35

McLeod, P. L. (1992). "An assessment of the experimental literature on electronic support of group work: Results of a meta-analysis." *Human-Computer Interaction* 7: 257-280. DOI: 10.1207/s15327051hci0703_1. 67

Meginbir, L., M Hebert, et al. (2004). "E-learning environment for health informatics education." *Healthcare Information Management & Communications Canada* 18(1): 22-24. 17

Meyer, E. T. (2009). Moving from small science to big science: Social and organizational impediments to large scale data sharing. *e-Research: Transformation in Scholarly Practice*. N. W. Jankowski. New York, Routledge: 147-159. 44

Michener, W. K. and R. B. Waide (2008). The evolution of collaboration in ecology: Lessons from the U.S. Long-Term Ecological Research Program. *Scientific Collaboration on the Internet*. G. M. Olson, A. Zimmerman and N. Bos. Cambridge, MA, MIT Press: 297-309. 2, 26

Millen, D. R., J. Feinberg, et al. (2006). *Dogear: Social bookmarking in the enterprise*. CHI '06, New York, ACM. DOI: 10.1145/1124772.1124792. 17, 65

Miller, R., M. Hobday, et al. (1995). "Innovation in complex systems industries: the case of flight simulation." *Industrial and Corporate Change* 4(2): 363-400. DOI: 10.1093/icc/4.2.363. 65

REFERENCES

Monk, A. F. (2009). *Common ground in electronically mediated conversation.* San Rafael, CA, Morgan & Claypool. DOI: 10.2200/S00154ED1V01Y200810HCI001. 39, 79, 82

Mosier, J. N. and S. G. Tammaro (1997). "When are group scheduling tools useful?" *Computer Supported Cooperative Work* 6: 53-70. DOI: 10.1023/A:1008684204655. 66

Muller, M. J., M. E. Raven, et al. (2003). *Introducing chat into business organizations: toward an instant messaging maturity model.* GROUP '03, New York, ACM. DOI: 10.1145/958160.958168. 59

Myers, D. G. and S. M. Smith (2011). *Exploring Social Psychology*, McGraw-Hill. 37

Myers, J. D. (2008). A national user facility that fits on your desk: The evolution of collaboratories at the Pacific Northwest National Laboratory. *Scientific Collaboration on the Internet.* G. M. Olson, A. Zimmerman and N. Bos. Cambridge, MA, MIT Press: 121-134. 1, 9, 27, 38, 71

Nardi, B. A., D. J. Shiano, et al. (2004). "Why we blog." *Communications of the ACM* 47(12): 41-46. DOI: 10.1145/1035134.1035163. 63

Nardi, B. A., S. Whittaker, et al. (2000). *Interaction and outeraction: Instant messaging in action.* CSCW 2000, New York, ACM Press. DOI: 10.1145/358916.358975. 59

Nisbett, R. (2003). *The geography of thought: How asians and westerners think differently ... and why.* New York, Free Press. 52

Noll, A. M. (1992). "Anatomy of a failure: Picturephone revisited." *Telecommunications Policy* 16(4): 307-316. DOI: 10.1016/0308-5961(92)90039-R. 62

Nunamaker, J. F., R. O. Briggs, et al. (1996/97). "Lessons from a dozen years of group support systems research: A discussion of lab and field findings." *Journal of Management Information Systems* 13(3): 163-207. 67

Nunamaker, J. F., A. R. Dennis, et al. (1991). "Electronic meeting systems." *Communications of the ACM* 34(7): 40-61. DOI: 10.1145/105783.105793. 67, 78

O'Leary, M. B. and J. N. Cummings (2007). "The spatial, temporal, and configurational characteristics of geographic dispersion in teams." *MIS Quarterly* 31(3): 433-452. 107

O'Leary, M. B. and M. Mortensen (2010). "Go (con)figure: Subgroups, imbalance, and isolates in geographically dispersed teams." *Organization Science* 21(1): 115-131. DOI: 10.1287/orsc.1090.0434. 49

Ohta, Y. and H. Tamura (1999). *Mixed reality: Merging real and virtual worlds.* New York, Springer-Verlag. 65

Okada, K., F. Maeda, et al. (1994). *Multiparty videoconferencing at virtual social distance: MAJIC design.* CSCW '94, New York, ACM. DOI: 10.1145/192844.193054. 62

REFERENCES 129

Olson, G. M. and D. Atkins (1990). Supporting collaboration with advanced multimedia electronic mail: The NSF EXPRES Project. *Intellectual teamwork: Social and technological foundations of cooperative work.* J. Galegher, R. E. Kraut and C. Egido. Hillsdale, NJ, Lawrence Erlbaum Associates: 429-451. xiii

Olson, G. M., T. L. Killeen, et al. (2008). The upper atmospheric research collaboratory and the space physics and aeronomy research collaboratory. *Scientific Collaboration on the Internet.* G. M. Olson, A. Zimmerman and N. Bos. Cambridge, MA, MIT Press: 153-169. 1, 9, 21, 27, 28, 40

Olson, G. M. and A. Luo (2007). Intra- and inter-cultural collaborations in science and engineering. *Intercultural Collaboration.* T. Ishida, S. R. Fussell and P. T. J. M. Vossen. Heidelberg, Springer Berlin. 4568: 249-259. 45

Olson, G. M., & Olson, J. S. (1991). User-centered design of collaboration technology. *Journal of Organization Computing* 1(1), 61-83. 39

Olson, G. M. and J. S. Olson (2000). "Distance matters." *Human–Computer Interaction* 15: 139-179. DOI: 10.1207/S15327051HCI1523_4. 3, 33, 39, 40, 53, 75

Olson, G. M., A. Zimmerman, et al., Eds. (2008). *Scientific collaboration on the Internet.* Cambridge, MA, MIT Press. DOI: 10.7551/mitpress/9780262151207.001.0001. 3

Olson, J. S., M. Ellisman, et al. (2008). The biomedical informatics research network. *Scientific Collaboration on the Internet.* G. M. Olson, A. Zimmerman and N. Bos. Cambridge, MA, MIT Press: 221-232. DOI:10.7551/mitpress/9780262151207.003.0013. 2, 13, 27, 40, 54, 71, 83

Olson, J. S., E. C. Hofer, et al. (2008). A theory of remote scientific collaboration. *Scientific collaboration on the Internet.* G. M. Olson, A. Zimmerman and N. Bos. Cambridge, MA, MIT Press: 73-97. DOI:10.7551/mitpress/9780262151207.003.0005. 33, 45, 102

Orlikowski, W. J. (1992) *Learning from notes: Organizational issues in groupware implementation.* CSCW 1992, New York, ACM Press. DOI: 10.1145/143457.143549. 14

Page, S. E. (2007). *The difference: How the power of diversity creates better groups, firms, schools, and societies.* Princeton, NJ, Princeton University Press. 2

Palen, L. and J. Grudin (2002). Discretionary adoption of group support software: Lessons from calendar applications. *Implementing collaboration technologies in industry.* B. E. Munkvold. London, Springer-Verlag: 159-180. 66

Palen, L., S. Vieweg, et al. (2011). "Supporting "everyday analysts" in time- and safety-critical situations." *The Information Society Journal* 27(1): 52-82. DOI: 10.1080/01972243.2011.534370. 2

REFERENCES

Palen, S. and S. B. Liu (2007). *Citizen communication in a crisis: Anticipating a future of ICT-supported public participation*. CHI 2007, ACM. DOI: 10.1145/1240624.1240736. 2

Paolacci, G., J. Chandler, et al. (2010). "Running experiments on Amazon Mechanical Turk." *Judgment and Decision Making* 5(5): 411-419. 27

Park, S. Y., S. Y. Lee, et al. (2012). "The effects of EMR deployment on doctors' work practices: A qualitative study in the emergency department of a teaching hospital." *International Journal of Medical Informatics* 81(3): 204-217. DOI: 10.1016/j.ijmedinf.2011.12.001. 13

Pascale, R. T. (1978). "Communication and decision making across cultures: Japanese and American comparisons." *Administrative Science Quarterly* 23(1): 91-110. DOI: 10.2307/2392435. 41

Patten, M. Q. (1990). *Qualitative evaluation and research methods*. Newbury Park, CA, Sage Publications. 7

Pevsner, J. (2009). *Bioinformatics and functional genomics*, Wiley-Blackwell. 1

Piper, T. and D. Vaver (2010). Ownership of medical images in eScience collaborations: Learning from the diagnostic mammography national database. *World Wide Research*. W. Dutton and P. W. Jeffreys. Cambridge, MA, MIT Press: 214-217. 54

Polanyi, M. (1961). "Knowing and being." *Mind N.S.* 70(280): 458-470. DOI: 10.1093/mind/LXX.280.458. 51

Porter, A. L. and I. Rafols (2009). "Is science becoming more interdisciplinary? Measuring and mapping six research fields over time." *Scientometrics* 81(3): 719-745. DOI: 10.1007/s11192-008-2197-2. 2

Ren, Y., R. Kraut, et al. (2007). "Applying common identity and bond theory to design of online communities." *Organizational Studies* 28(3): 377-408. DOI: 10.1177/0170840607076007. 72

Rheingold, H. (2002). *Smart mobs*. New York, Basic Books. 14

Ribes, D. and G. C. Bowker (2008). Organizing for multidisciplinary collaboration: The case of the geosciences network. *Scientific collaboration on the Internet*. G. M. Olson, A. Zimmerman and N. Bos. Cambridge, MA, MIT Press: 311-330. 44, 45

Rocco, E. (1998). *Trust breaks down in electronic contexts but can be repaired by some initial face-to-face contact*. CHI 1998, ACM. DOI: 10.1145/274644.274711. 47

Rosen, E. (2009). *The culture of collaboration: Maximizing time, talent and tools to create value in the global economy*. San Francisco, Red Ape Publishing. 3

Ross, C., M. Terras, et al. (2010). "Enabled backchannel: conference Twitter use by digital humanists." *Journal of Documentation* 67(2): 214-237. DOI: 10.1108/00220411111109449. 63

REFERENCES 131

Rousseau, D. M., S. B. Sitkin, et al. (1998). "Not so different after all: A cross-discipline view of trust." *Academy of Management Review* 23(3): 393-404. DOI: 10.5465/AMR.1998.926617. 46

Salvucci, D. D. and N. A. Taatgen (2011). *The multitasking mind*. New York, Oxford University Press. 60

Sarma, A., D. Redmiles, et al. (2010). "Categorizing the spectrum of coordination technology." *IEEE Computer* 43(6): 61-67. DOI: 10.1109/MC.2010.163. 57

Satzinger, J. and L. Olfman (1992). *A research program to assess user perceptions of group work support*. CHI '92, New York, ACM Press. DOI: 10.1145/142750.142765. 58

Sawyer, K. (2007). *Group genius: The creative power of collaboration*. New York, Basic Books. 3

Schmidt, K. (2002). "The problem with 'awareness': Introductory remarks on 'Awareness in CSCW.'." *Computer Supported Cooperative Work* 11: 285-298. DOI: 10.1023/A:1021272909573. 66

Semaan, B. and G. Mark (2012). *"Facebooking" towards crisis recovery and beyond: Disruption as an opportunity*. CSCW '12, New York, ACM Press. DOI: 10.1145/2145204.2145214. 77

Sewell, D. N. (2004). What's in a name? Defining electronic community. *Electronic Collaboration in the Humanities*. J. A. Inman, C. Reed and P. Sands. Mahwah, NJ, Lawrence Erlbaum Associates: 227-239. 17

Shirky, C. (2008). *Here Comes Everybody: The power of organizing without organizations*. New York, Penguin Press. 14

Shrum, W., I. Chompalov, et al. (2001). "Conflict and performance in scientific collaborations." *Social Studies of Science* 31(5): 681-730. DOI: 10.1177/030631201031005002. 45, 46

Sonderegger, P. (2009). Creating shared understanding across distance. *e-Research: Transformation in Scholarly Practice*. N. W. Jankowski. New York, Routledge: 129-146. 37, 40, 46, 47, 52

Sonnenwald, D. H. (2007). Scientific collaboration: A synthesis of challenges and strategies. *Annual review of information science and technology*. G. Cronin. Nedford, NJ, Information Today: 643-681. DOI: 10.1002/aris.2007.1440410121. 53

Spencer, B. F., Jr., R. Butler, et al. (2008). NEESgrid: Lessons learned for future cyberinfrastructure development. *Scientific collaboration on the Internet*. G. M. Olson, A. Zimmerman and N. Bos. Cambridge, MA, MIT Press: 331-347. DOI: 10.7551/mitpress/9780262151207.003.0019. 1, 12, 37

Sproull, L. and S. Kiesler (1991). *Connections: New ways of working in the neetworked organization*. Cambridge, MA, MIT Press. 58, 60

132 REFERENCES

Sproull, L. and S. Kiesler (2005). Public volunteer work on the Internet. *Transforming enterprise: The economic and social implications of information technology.* D. W.H., B. Kahin, R. O'Callaghan and A. W. Wyckoff. Cambridge, MA, MIT Press: 361-374. 14

Starbird, K. and L. Palen (2011). *"Voluntweeters":Self-organizing by digital volunteers in times of crisis.* CHI 2011. New York, ACM: 1071-2080. DOI: 10.1145/1978942.1979102. 64

Steinberg, S. G. (1996). Netheads vs Bellheads. *Wired.* 4. 39

Steiner, I. D. (1972). *Group processes and productivity.* New York, Academic Press. 35

Stipelman, B., A. Feng, et al. (2010). "The relationship between collaborative readiness and scientific productivity in the Transdisciplinary Research on Energetics and Cancer (TREC) Centers." *Annals of Behavioral Medicine* 39(Supplement 1): S143. 46

Stokes, D. E. (1997). *Pasteur's quadrant: Basic science and technological innovation.* Washington, DC, Brookings Institution Press. 25, 29

Stokols, D., J. Fuqua, et al. (2003). "Evaluating transdisciplinary science." *Nicotine & Tobacco Research 5(Supplement 1)*: S21-S39. DOI: 10.1080/14622200310001625555. 53

Stokols, D., K. L. Hall, et al. (2010). Cross-disciplinary team science initiatives: Research, training, and translation. *Oxford handbook on interdisciplinarity.* R. Frodeman, J. T. Klein and C. Mitcham. New York, Oxford University Press: 471-493. 2

Stokols, D., K. L. Hall, et al. (2008). "The science of team science: Overview of the field and introduction to the supplement." *American Journal of Preventive Medicine* 25(2S): S77-S89. DOI: 10.1016/j.amepre.2008.05.002. 3, 26, 33, 43

Stokols, D., K. L. Hall, et al. (2013). Transdisciplinary public health: Core characteristics, definitions, and strategies for success. *Transdisciplinary public health: Research, methods, and practice.* D. Haire-Joshu and T. D. McBride. San Francisco, Jossey-Bass: 2-30. 21

Stokols, D., R. Harvey, et al. (2005). "In vivo studies of transdisciplinary scientific collaboration: Lessons learned and implications for active living research." *American Journal of Preventive Medicine* 28(Supplement 2): 202-213. DOI: 10.1016/j.amepre.2004.10.016. 53

Surowiecki, J. (2005). *The wisdom of crowds.* New York, Anchor Books. 14, 15, 75

Szpir, M. (2002). "Clickworkers on Mars." *American Scientist* 90(3): 226. 15

Takeuchi, K., J. C. Lin, et al. (2010). "Scheduling with package auctions." *Experimental economics* 13: 476-499. DOI: 10.1007/s10683-010-9252-6. 70

Tanenbaum, A. S. and D. J. Wetherall (2011). *Computer networks.* Upper Saddle River, NJ, Prentice Hall. 12

REFERENCES 133

Tang, J. C., J. Marlow, et al. (2012). "Time travel proxy: Using lightweight video recordings to create asynchronous, interactive meetings." *Proceedings of the ACM Conference on Human Factors in Computing Systems*, CHI 2012. : 3111-3120. DOI: 10.1145/2207676.2208725. 52

Teasley, S. D. (2000). Conversation with J. S. Olson. 28

Teasley, S. D., T. Schleyer, et al. (2008). Three distributed biomedical research centers. *Scientific collaboration on the Internet*. G. M. Olson, A. Zimmerman and N. Bos. Cambridge, MA, MIT Press: 233-250. DOI: 10.7551/mitpress/9780262151207.003.0014. 44

Thompson, J. (1967). *Organizations in action*. New York, McGraw-Hill. 35

Thompson, K. (2008). *The networked enterprise: Competing for the future through virtual enterprise networks*. Tampa, FL, Meghan-Kiffer Press. 8

Turkle, S. (1995). *Life on the screen: Identity in the age of the Internet*. New York, Simon & Schuster. 64

Tyran, K. L., C. K. Tyran, et al. (2003). Exploring emerging leadership in virtual teams. *Virtual teams that work: Creating conditions for virtual team effectiveness*. C. B. Gibson and S. B. Cohen. San Francisco, Jossey-Bass: 183-195. 50, 51

Ulijn J. M and X. Li (1995). "Is interrupting impolite? Some temporal aspects of turn taking in Chinese-Western and other intercultural business encounters." *Text* 15(4): 589-627. DOI: 10.1515/text.1.1995.15.4.589. 61

Veinott, E. S., J. S. Olson, et al. (1999). *Video helps remote work: Speakers who need to negotiate common ground benefit from seeing each other*. CHI '99, New York, ACM Press. DOI: 10.1145/302979.303067. 41, 80

Vertegaal, R., I. Weevers, et al. (2003). *GAZE-2: Conveying eye contacdt in group video conferencing using eye-controlled camera direction*. CHI '03, New York, ACM. DOI: 10.1145/642611.642702. 62

Vieweg, S., A. L. Hughes, et al. (2010). *Microblogging during two natural hazards events: What Twitter may contribute to situational awareness*. CHI 2010. New York, ACM: 1079-1088. DOI: 10.1145/1753326.1753486. 64

Voida, A., M. E. Harmon, et al. (2011). *Homebrew databases: Complexities of everyday information management in nonprofit organizations*. CHI 2011, New York, ACM Press. DOI: 10.1145/1978942.1979078. 8, 13

Voida, A., J. S. Olson, et al. (2013). *Turbulence in the clouds: Challenges of cloud-based information work*. CHI 2013. New York, ACM. DOI: 10.1145/2470654.2481313. 13, 54. 71, 73 von Ahn, L., M. Blum, et al. (2004). "Telling humans and computers apart automatically." *Communications of the ACM* 47: 57-60. DOI: 10.1145/966389.966390. 16

von Ahn, L. and L. Dabbish (2004). *Labeling images with a computer game*. CHI 2004, New York, ACM Press. DOI: 10.1145/985692.985733. 16

von Ahn, L., B. Maurer, et al. (2008). "reCAPTCHA: Human-based character recognition via web security measures." *Science* 321: 1465-1468. DOI: 10.1126/science.1160379. 16

Walther, J. B. and U. Bunz (2005). "The rules of virtual groups: Trust, liking and performance in computer-mediated communication." *Journal of Communication* 55(4): 828-846. DOI: 10.1111/j.1460-2466.2005.tb03025.x. 51

Wankel, C. and R. Hinrichs (2011). *Transforming virtual world learning*. Bingley, UK, Emerald Group Publishing. 65

Ware, C. (2013). *Information visualization: Perception for fesign*. Waltham, MA, Morgan Kaufmann. 13

Weisband, S. (2008). Research challenges in studying leadership at a distance. *Leadership at a distance: Research in technologically-supported work*. S. Weisband. New York, Lawrence Erlbaum Associates: 3-11. 50

Whittaker, S. (2011). Personal communication. 65

Whittaker, S., V. Bellotti, et al. (2005). "Introduction to this special issue on revisiting and reinventing e-mail." *Human-Computer Interaction* 20(1): 1-9. DOI: 10.1207/s15327051hci2001&2_1. 58

Whittaker, S. and C. Sidner (1996). *E-mail overload: Exploring personal information management of e-mail*. CHI '96. New York, ACM Press: 276-283. DOI: 10.1145/238386.238530. 58

Wiggins, A. and K. Crowston (2010). "Developing a conceptual model of virtual organizations for citizen science." *International Journal of Organizational Design and Engineering* 1(1/2): 148-162. DOI: 10.1504/IJODE.2010.035191. 15

Williams, H. M., S. K. Parker, et al. (2007). "Perceived dissimilarity and perspective taking within work teams." *Group & Organization Management* 32(5): 569-597. DOI: 10.1177/1059601106293769. 40

Woolley, A. W., C. F. Chabris, et al. (2010). "Evidence for a collective intelligence factor in the performance of human groups." *Science* 29: 686-688. DOI: 10.1126/science.1193147. 43, 44

Wuchty, S., B. F. Jones, et al. (2007). "The increasing dominance of teams in production of knowledge." *Science* 316: 1036-1039. DOI: 10.1126/science.1136099. 2

Wulf, W. A. (1993). "The collaboratory opportunity." *Science* 261(5123): 854-855. DOI: 10.1126/science.8346438. 1, 87

Yuki, M., W. W. Maddux, et al. (2005). "Cross-cultural differences in relationship- and group-based trust." *Personality and Social Psychology Bulletin* 31: 48-61. DOI: 10.1177/0146167204271305. 47

Zaccaro, S. J., M. A. Marks, et al., Eds. (2013). *Multiteam systems: An organizational form for dynamic and complex environments*. Organization and Management Series. New York, Routledge. 21, 49

Zhang, J., Y. Qu, et al. (2010). *A case study of micro-blogging in the enterprise: Use, value, and related issues*. CHI 2010. New York, ACM: 123-132. DOI: 10.1145/1753326.1753346. 64

Zhao, D. and M. B. Rosson (2009). *How and why people Twitter: The role that micro-blogging plays in informal communication at work*. GROUP 2009. New York, ACM: 243-252. DOI: 10.1145/1531674.1531710. 64

Zheng, J., B. Veinott, et al. (2002). *Trust without touch: Jumpstarting long-distance trust with initial social activities*. CHI 2002, New York, ACM Press. DOI: 10.1145/503376.503402. 47

Zigurs, I. and B. K. Buckland (1998). "A theory of task/technology fit and group support systems effectiveness." *MIS Quarterly* 22(3): 313-334. DOI: 10.2307/249668. 80

Zimmerman, A. S. (2008). "New knowledge from old data: The role of standards in the sharing and reuse of ecological data." *Science, Technology & Human Values* 33(5): 631-652. DOI: 10.1177/0162243907306704. 13

Author Biographies

Judith Olson is the Bren Professor of Information and Computer Sciences in the Informatics Department at the UC Irvine, with courtesy appointments in the School of Social Ecology and the Merage School of Business.

She has researched teams whose members are not collocated for over 20 years, summaries of which are found in her most cited paper, "Distance Matters," (Olson & Olson, 2000), and in her key theoretical contribution in the book *Scientific Collaboration on the Internet* (Olson, Zimerman, and Bos, Eds., 2008).

Her current work focuses on ways to verify the theory's components while at the same time helping new scientific collaborations succeed. She has studied distributed teams both in the field and in the laboratory, the latter focusing on the communication hurdles distributed teams have and the consequent underutilization of remote team members skills and the reduction in trust.

She is a Fellow of the Association for Computing Machinery and with her husband and colleague, Gary Olson, holds the Lifetime Achievement award from the Special Interest Group in Computer Human Interaction.

Gary Olson is a professor at the University of California at Irvine, in the Donald Bren School of Information and Computer Sciences. He is also a Professor Emeritus at the University of Michigan.

His research spans the fields of Human-Computer Interaction and Computer-Supported Cooperative Work, with particular interests in collaborative technologies and their role in supporting long-distance work. His research interests include cognition, problem-solving, reasoning, and communication in social and physical settings.

Olson is a member of the Association of Computing Machinery, the Association for Psychological Science, and the American Psychological Association. He is currently a member of the editorial boards of the *Journal of Organizational Computing*, *Human-Computer Interaction*, *Computer Supported Cooperative Work: An International Journal*, the *Journal of Biomedical Discovery and Collaboration*, and *Foundations and Trends in Human-Computer Interaction*.

Olson is a Fellow of the Association of Computing Machinery, the Association for Psychological Science, and the American Psychological Association. In 2003, he was elected to the CHI Academy. In 2006, he received the CHI Lifetime Achievement Award along with his wife, Dr. Judith S. Olson.

CPSIA information can be obtained
at www.ICGtesting.com
Printed in the USA
LVOW02s0012291015

460180LV00002B/12/P

9 781608 450503